D1645539

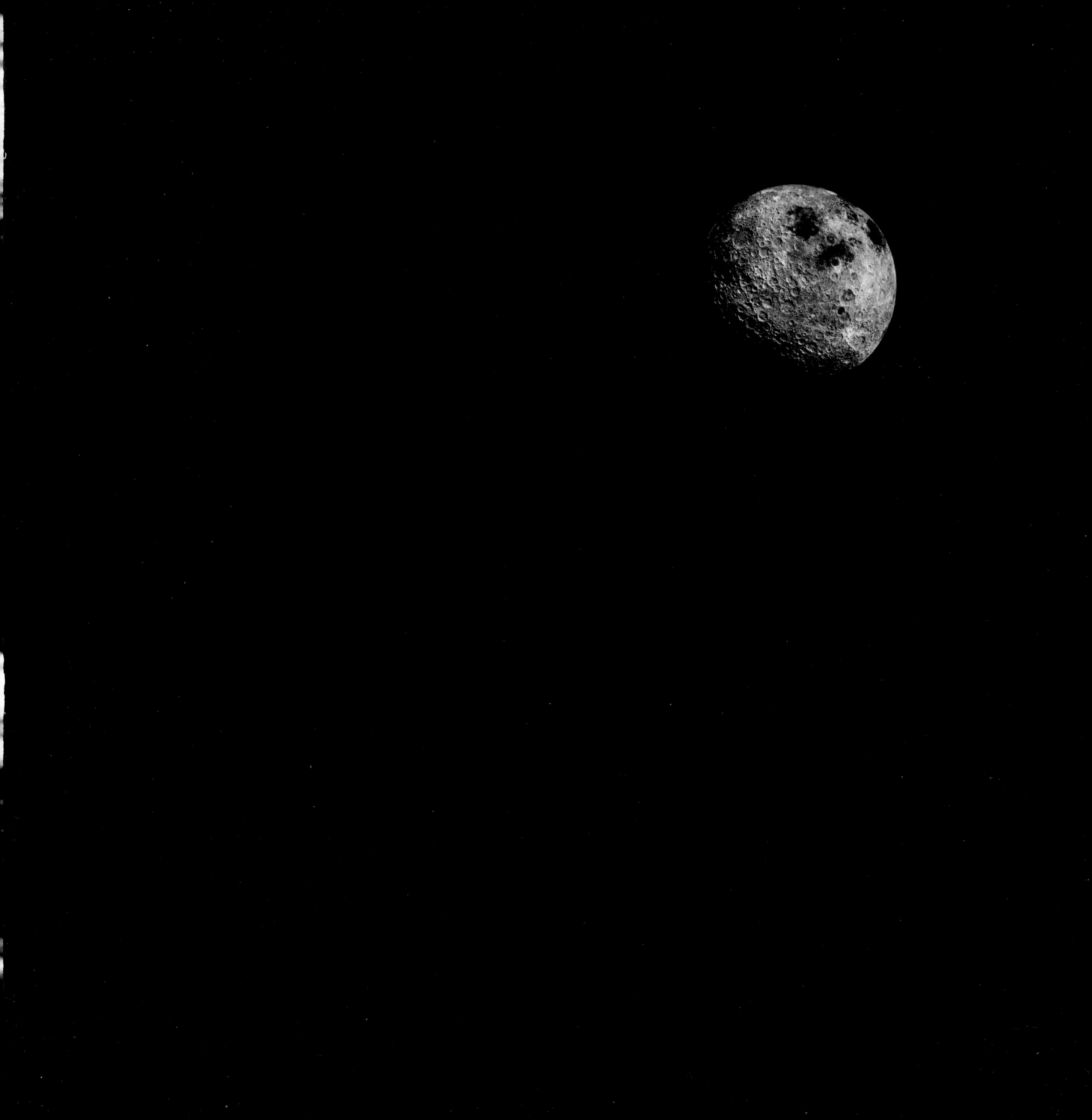

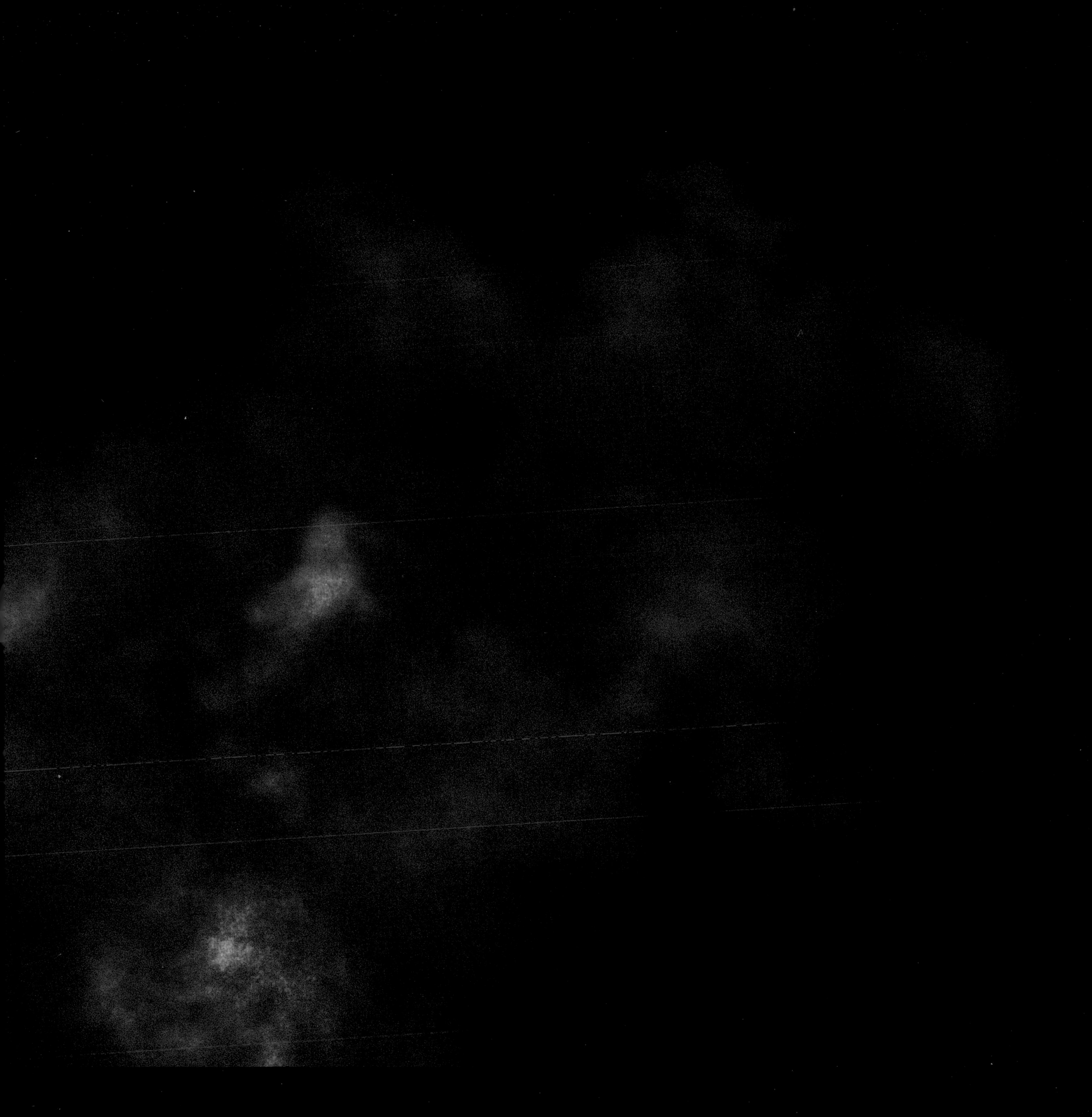

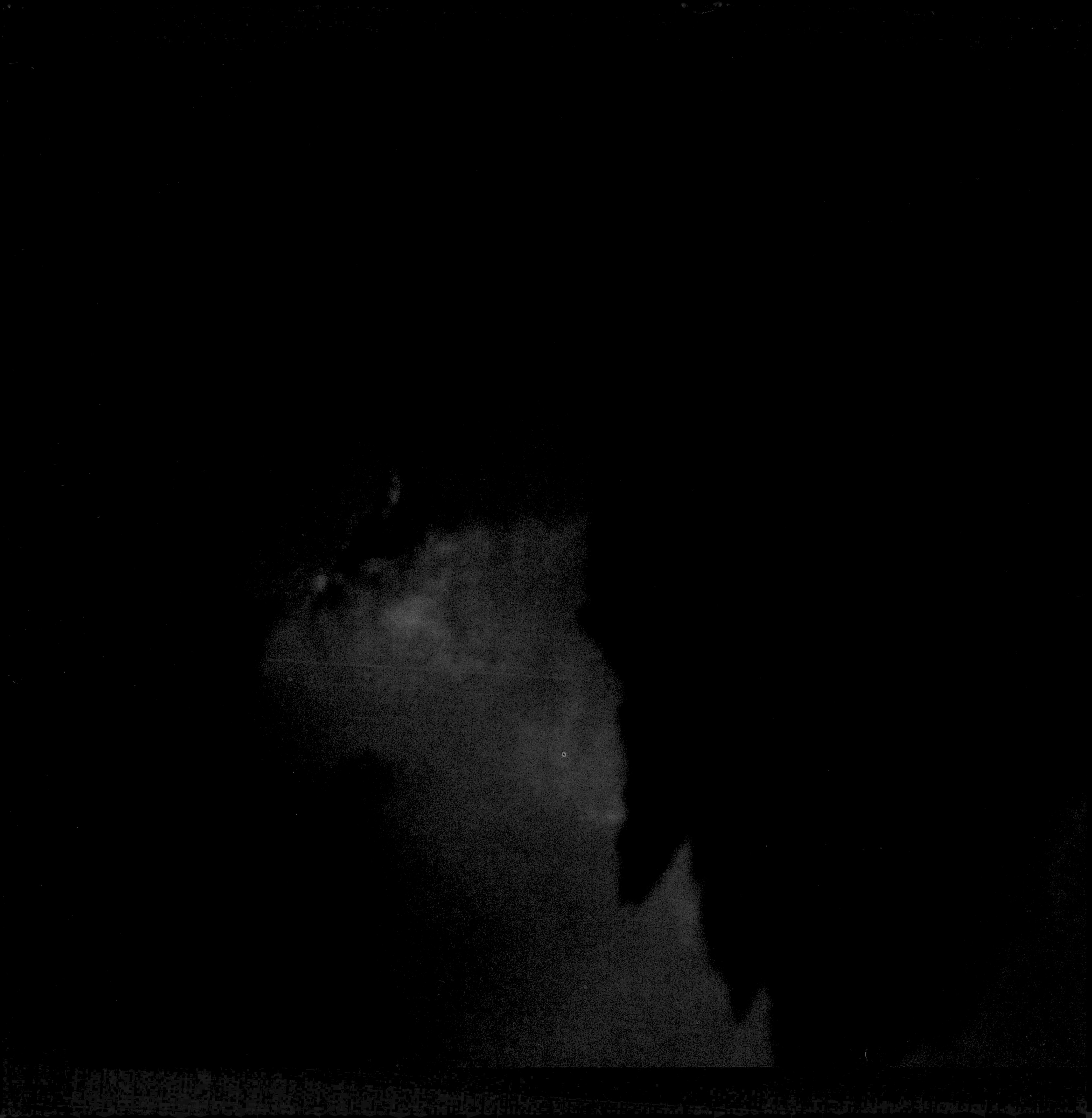

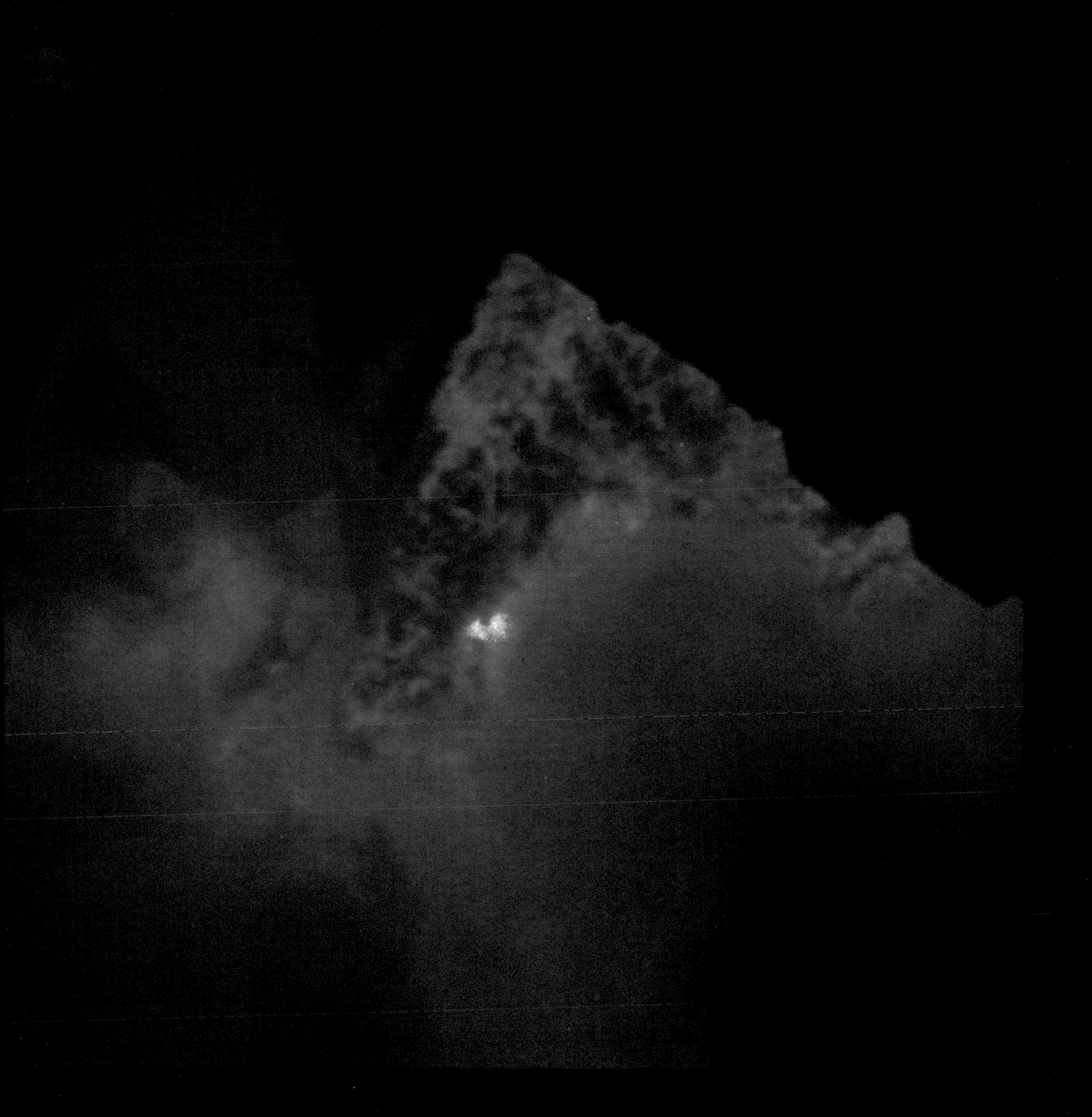

MICHAEL LIGHT

ESSAY BY ANDREW CHAIKIN

EDWIN E. ALDRIN, JR.

CHARLES CONRAD, JR.

WILLIAM A. ANDERS

R. WALTER CUNNINGHAM

NEIL A. ARMSTRONG

CHARLES M. DUKE, JR.

ALAN L. BEAN

DONN F. EISELE

FRANK BORMAN

RONALD E. EVANS

EUGENE A. CERNAN

RICHARD F. GORDON, JR.

ROGER B. CHAFFEE

VIRGIL I. GRISSOM

MICHAEL COLLINS

FRED W. HAISE, JR.

APOLLO 1 APOLLO 7 APOLLO 8 APOLLO 9 APOLLO 10 APOLLO 11

JAMES B. IRWIN

JAMES A. LOVELL, JR.

T. KENNETH MATTINGLY II

JAMES A. McDIVITT

EDGAR D. MITCHELL

STUART A. ROOSA

WALTER M. SCHIRRA, JR.

HARRISON H. SCHMITT

RUSSELL L. SCHWEICKART

DAVID R. SCOTT

ALAN B. SHEPARD, JR.

THOMAS P. STAFFORD

JOHN L. SWIGERT, JR.

EDWARD H. WHITE II

ALFRED M. WORDEN

JOHN W. YOUNG

APOLLO 12 APOLLO 13 APOLLO 14 APOLLO 15 APOLLO 16 APOLLO 17

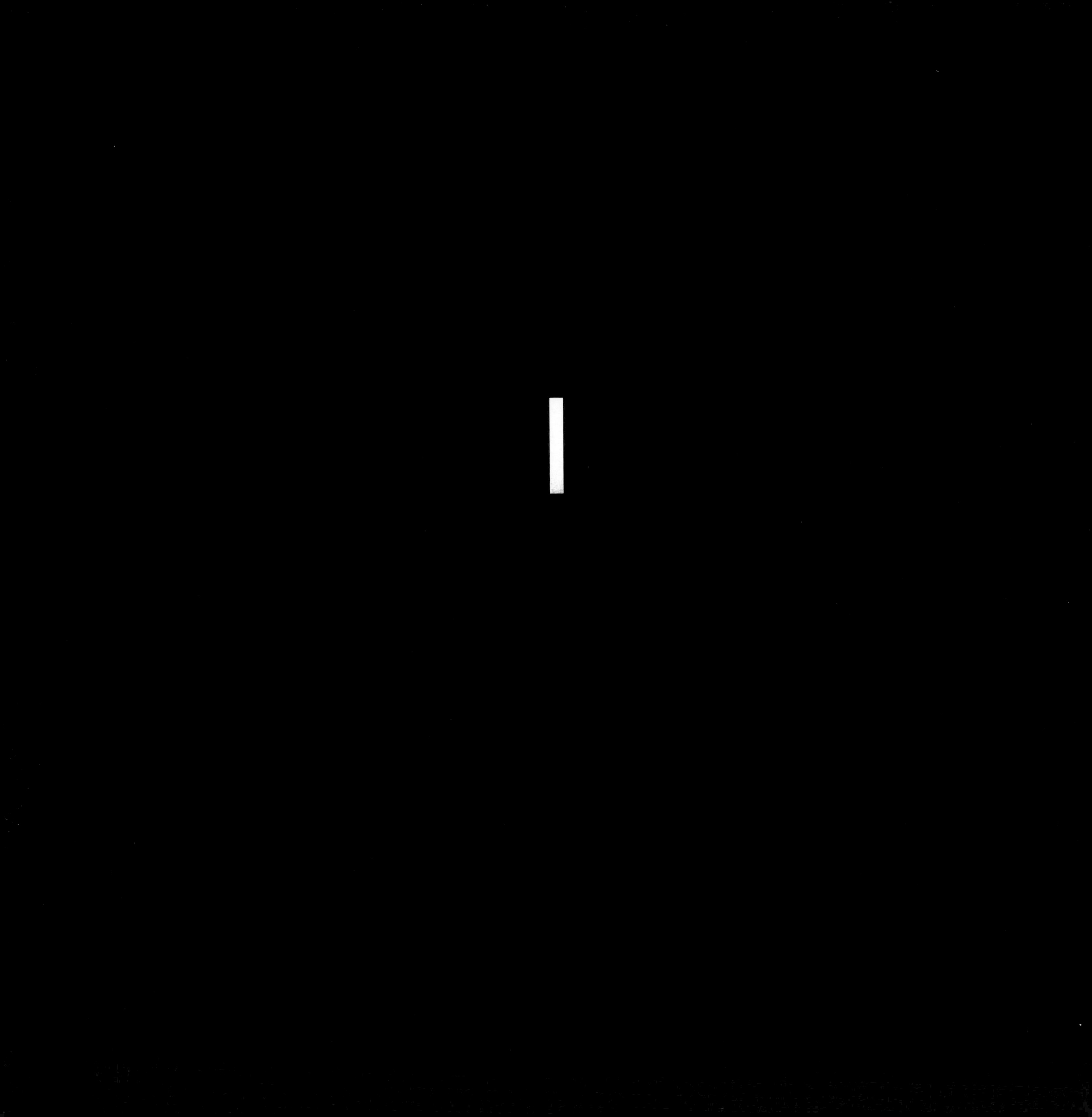

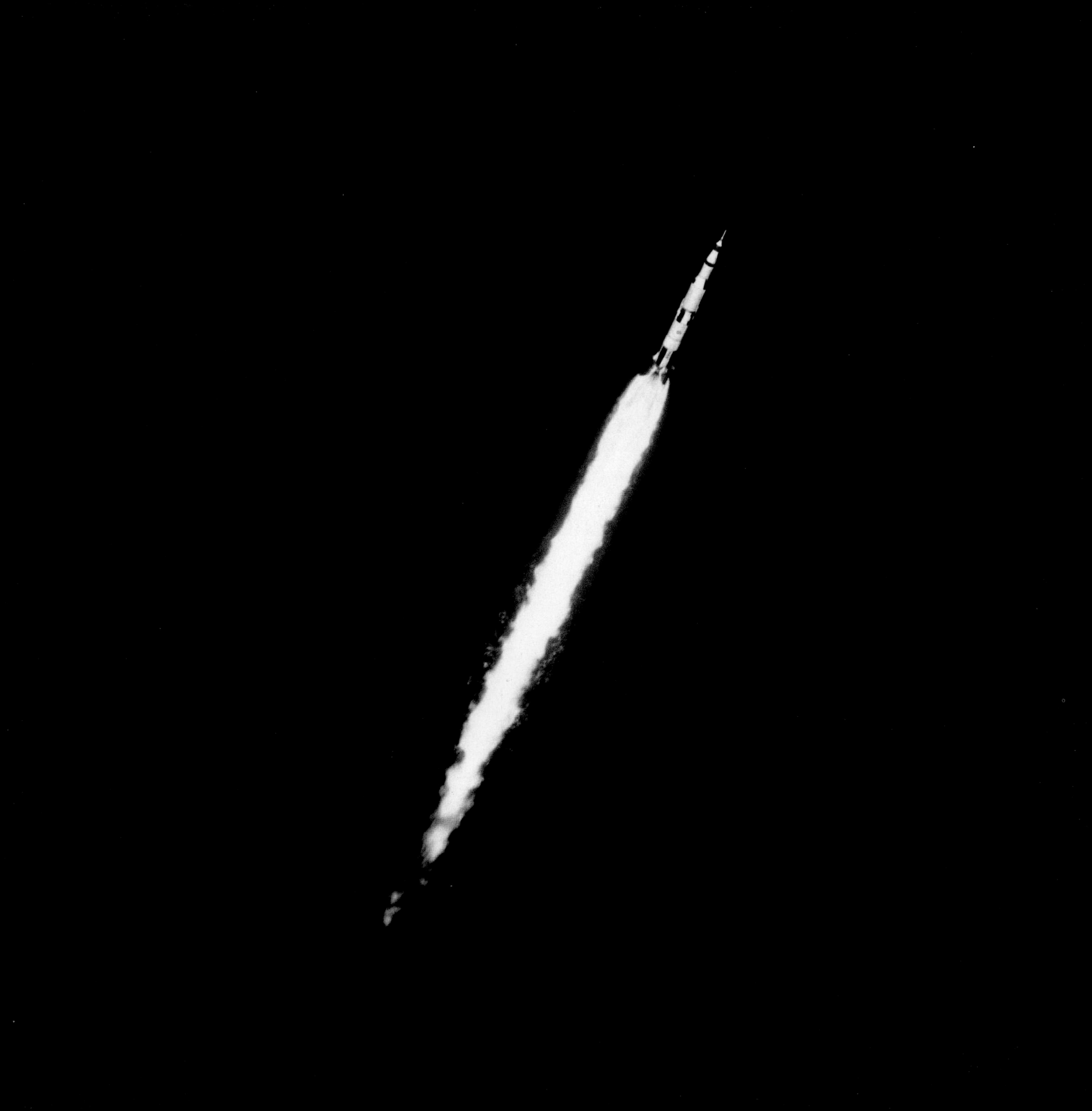

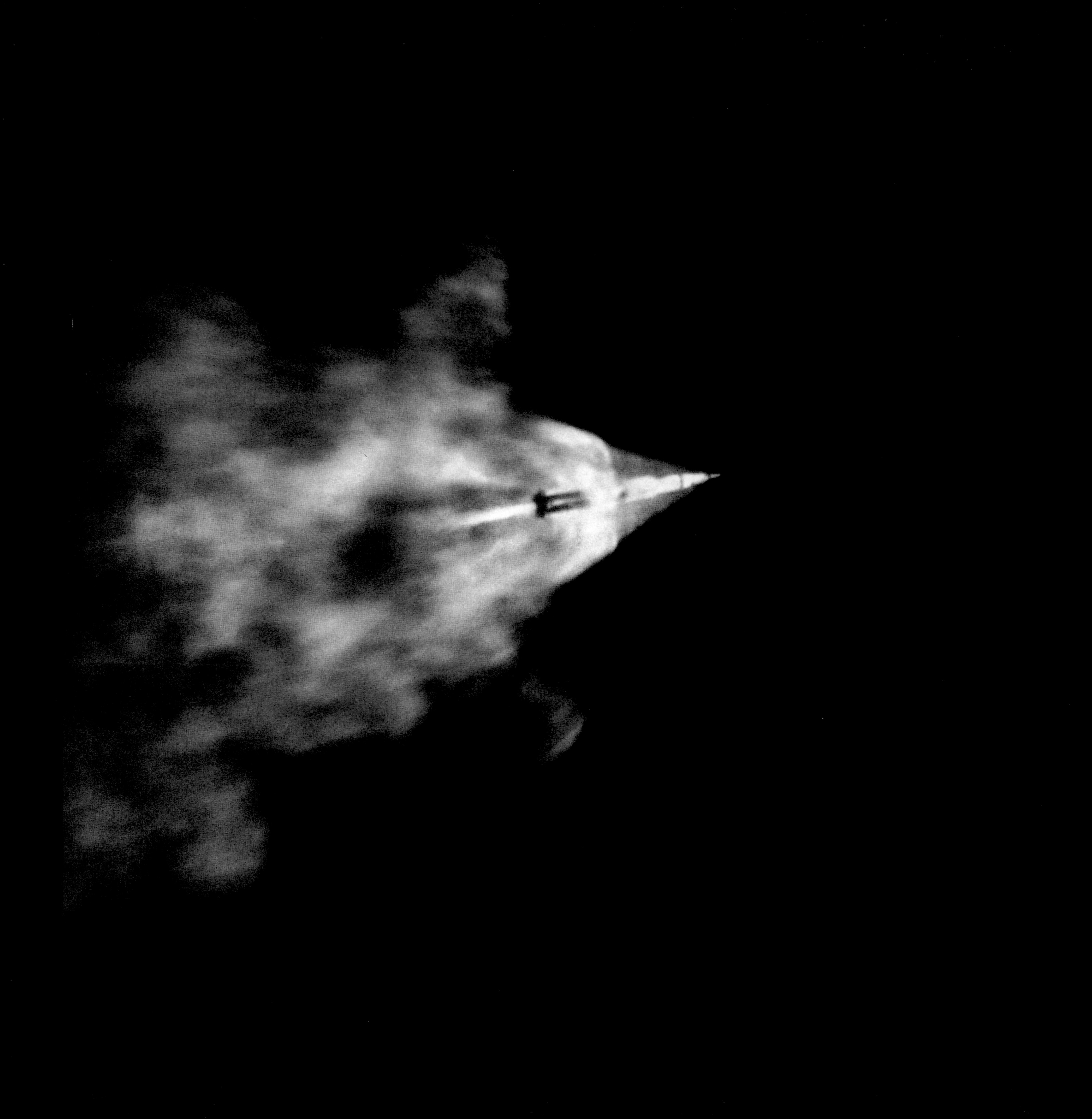

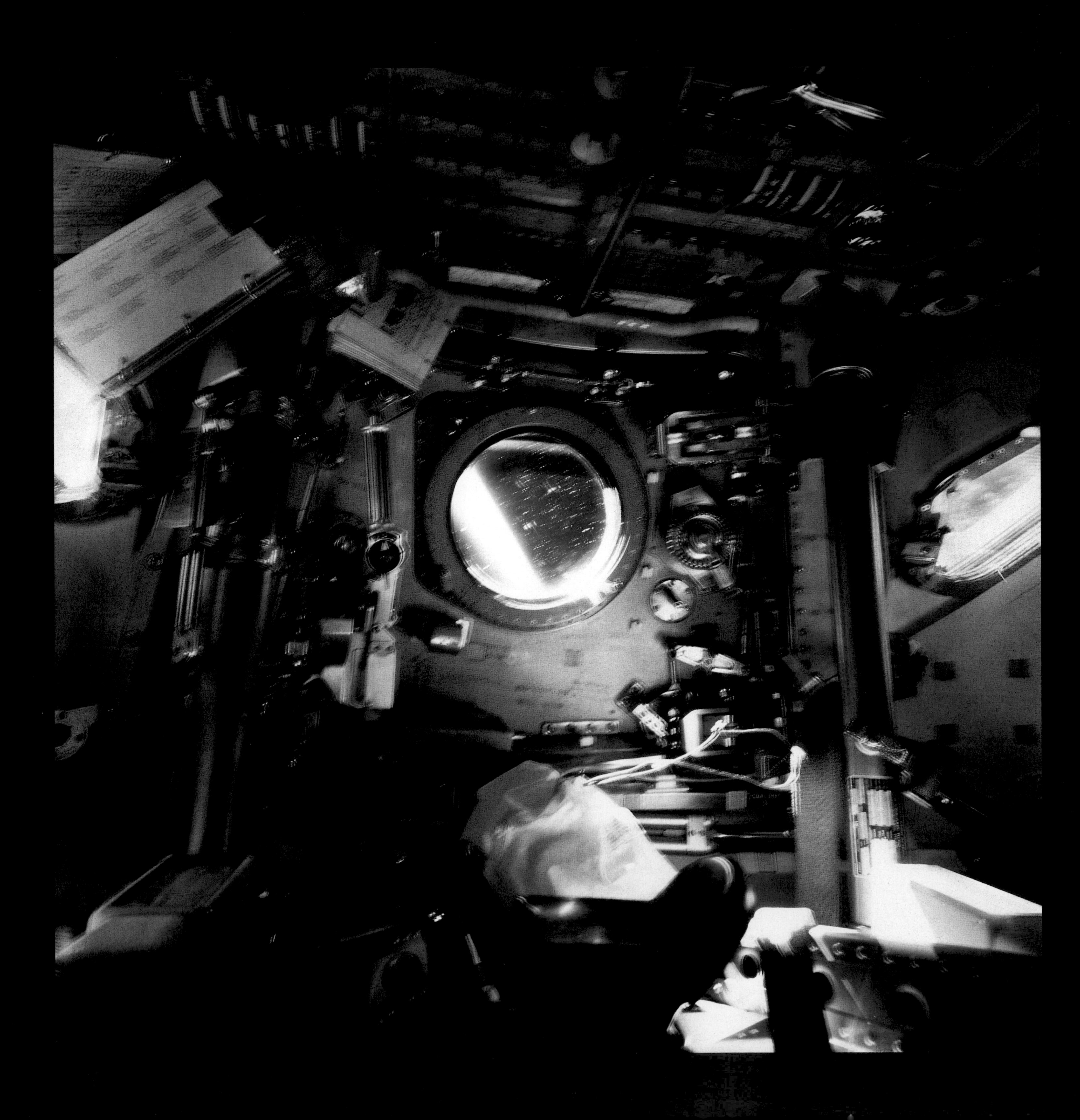

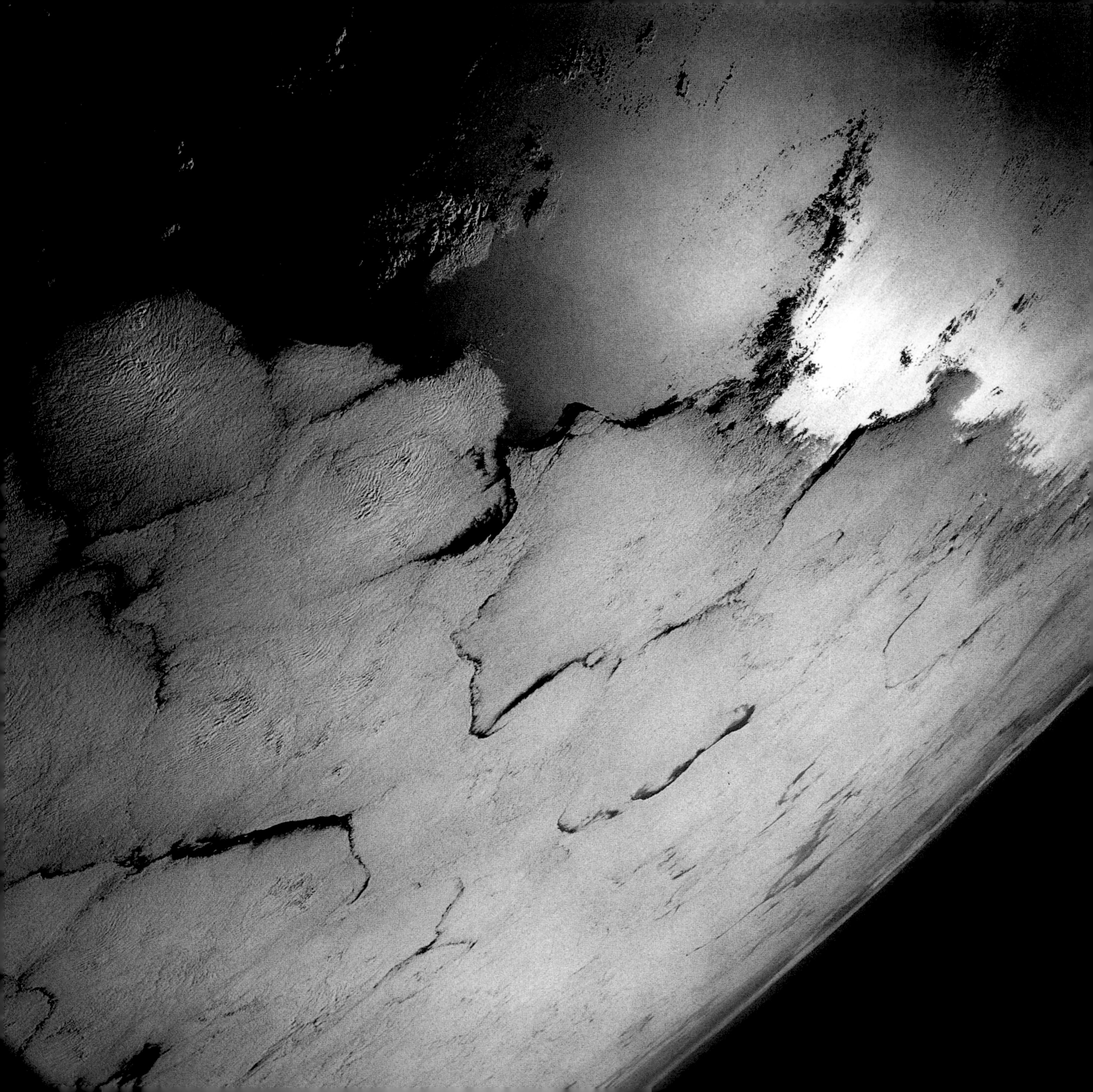

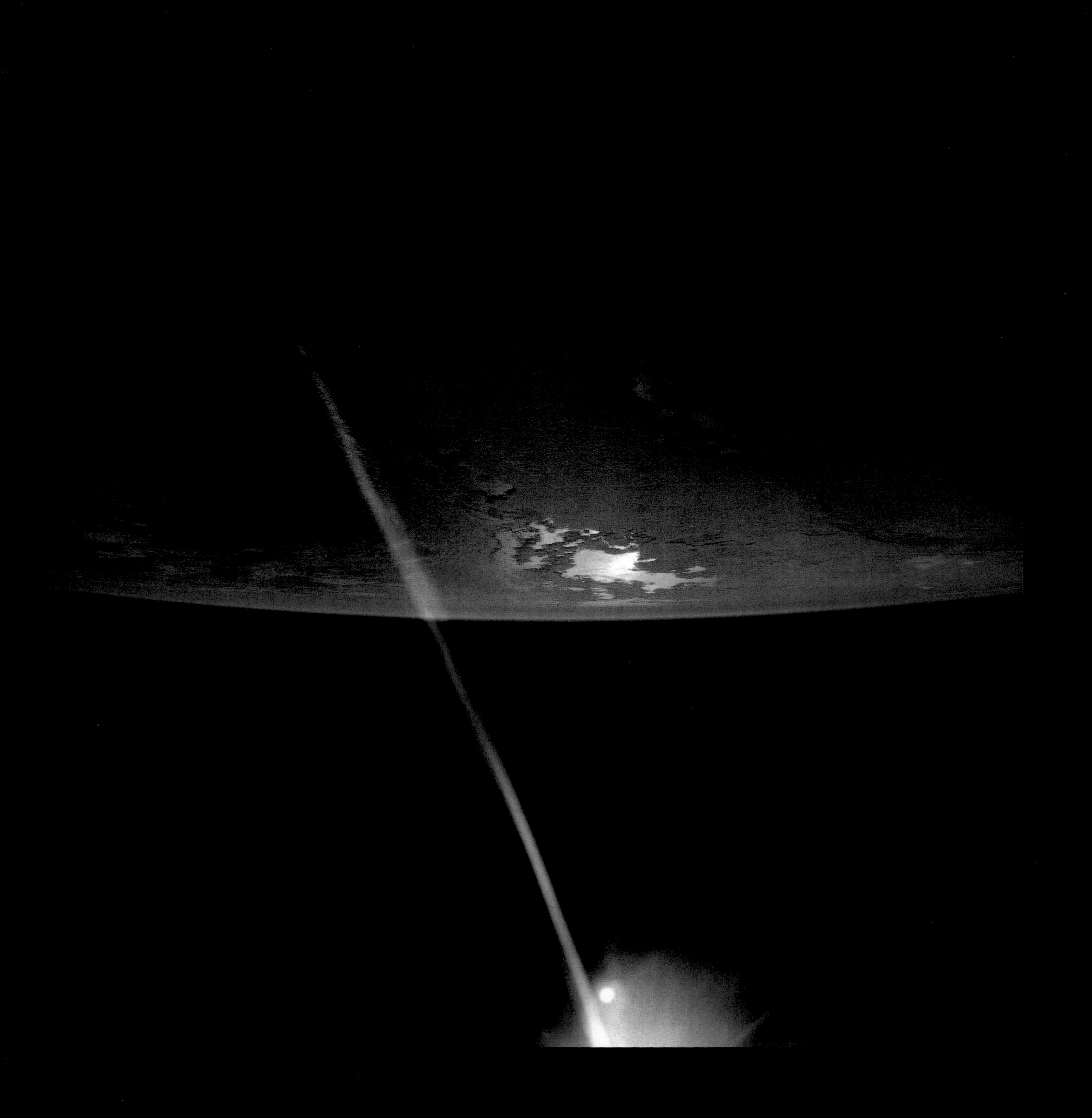

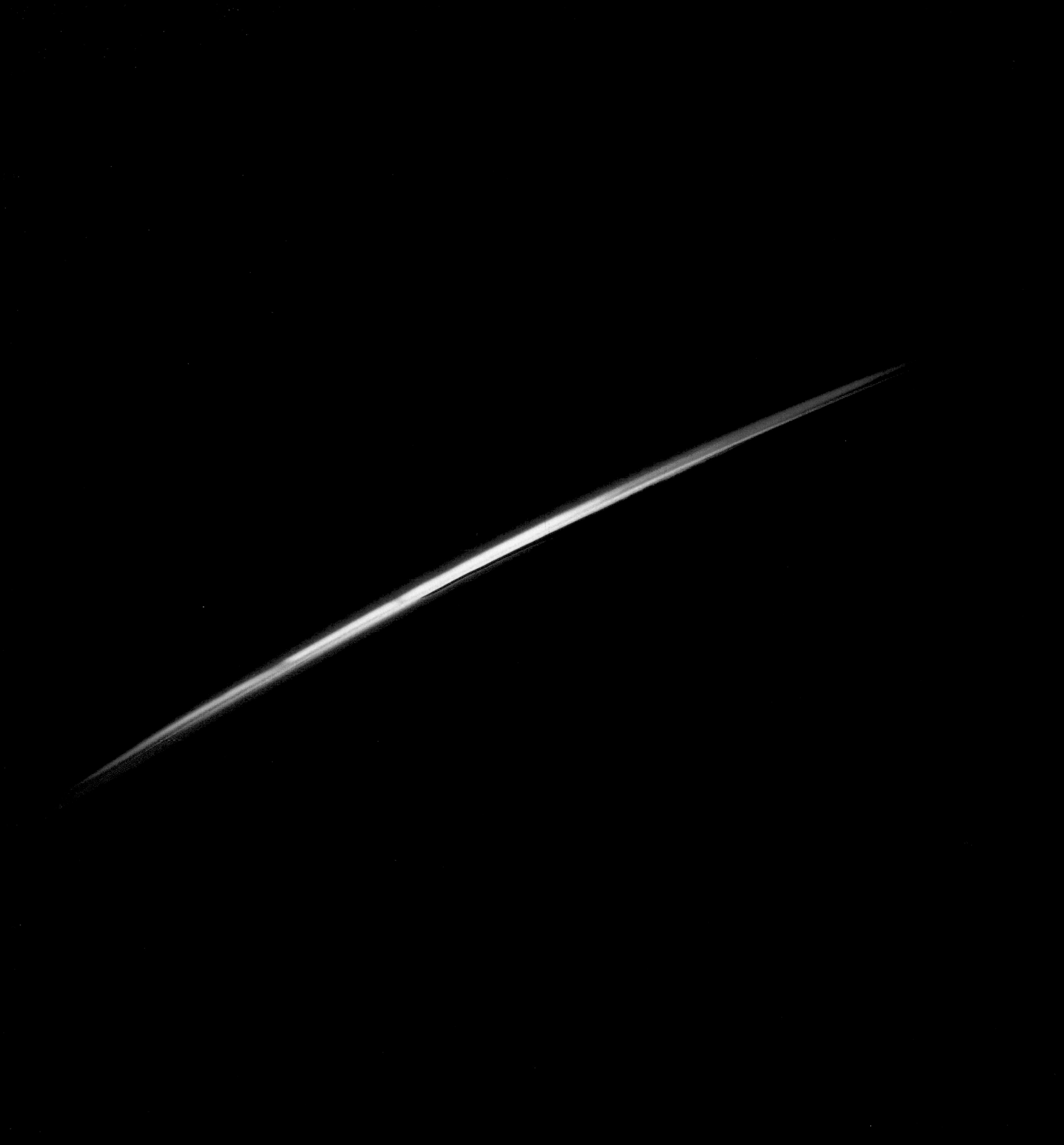

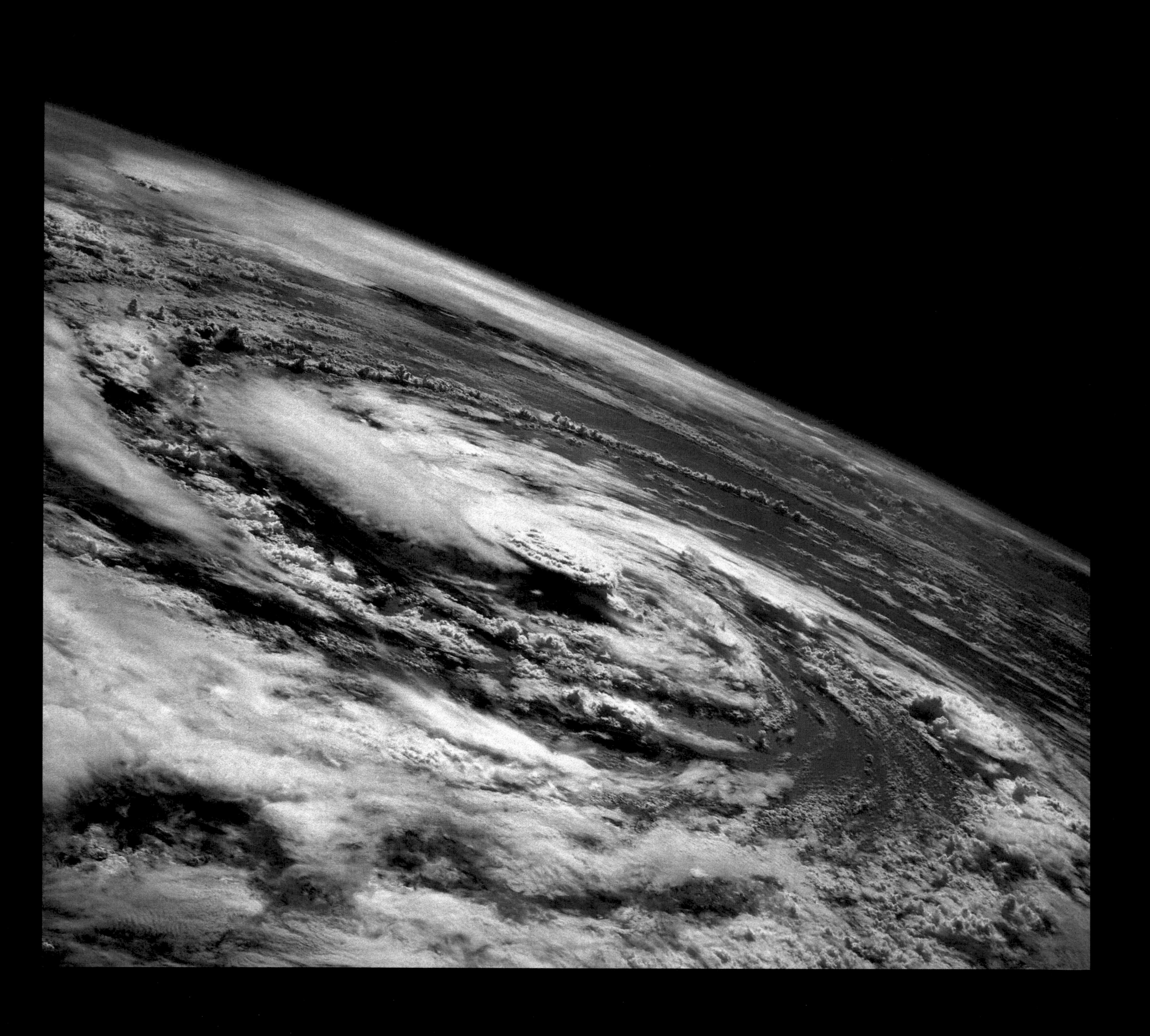

SCHIRRA
W. SCHIRRA

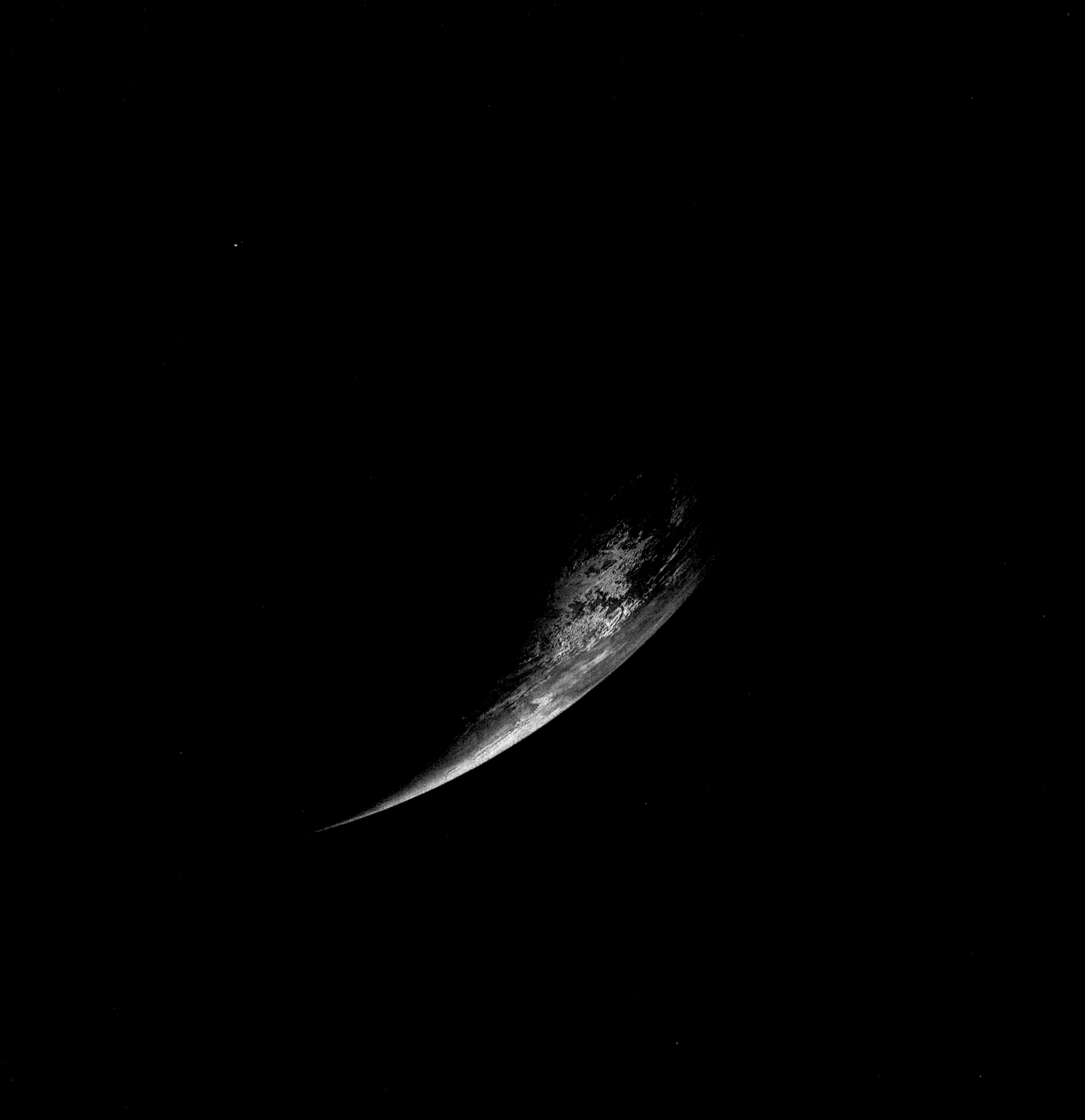

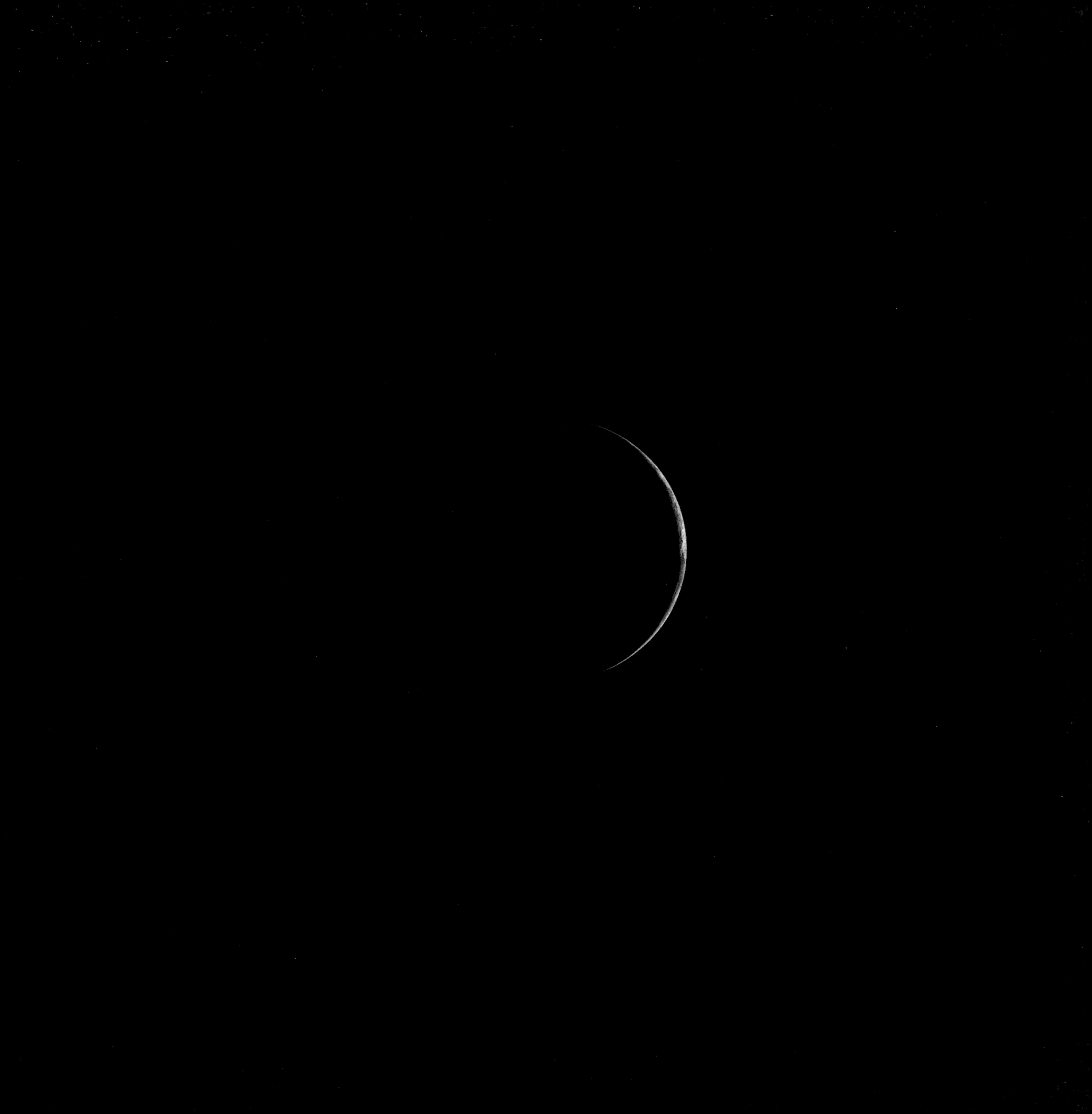

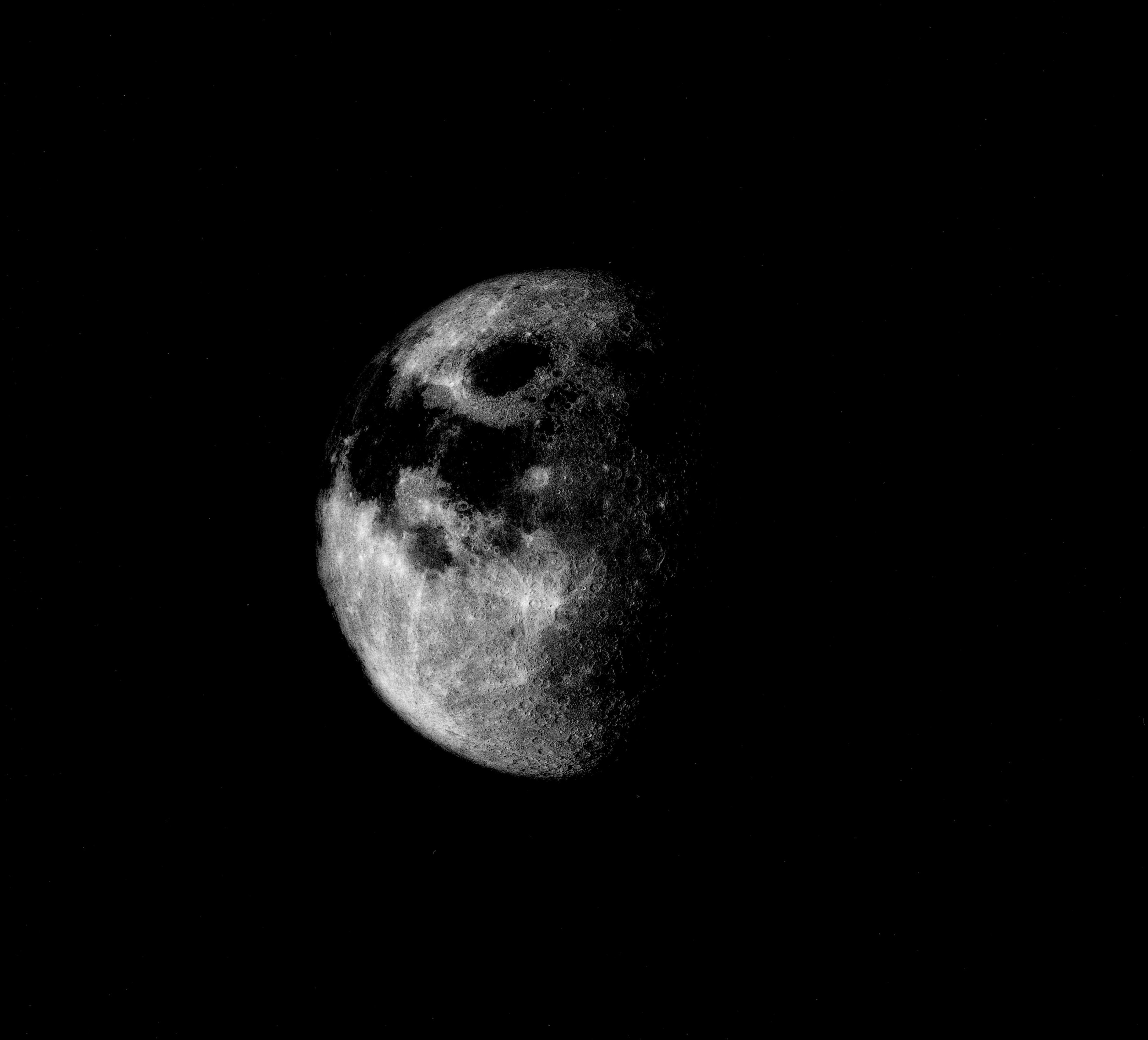

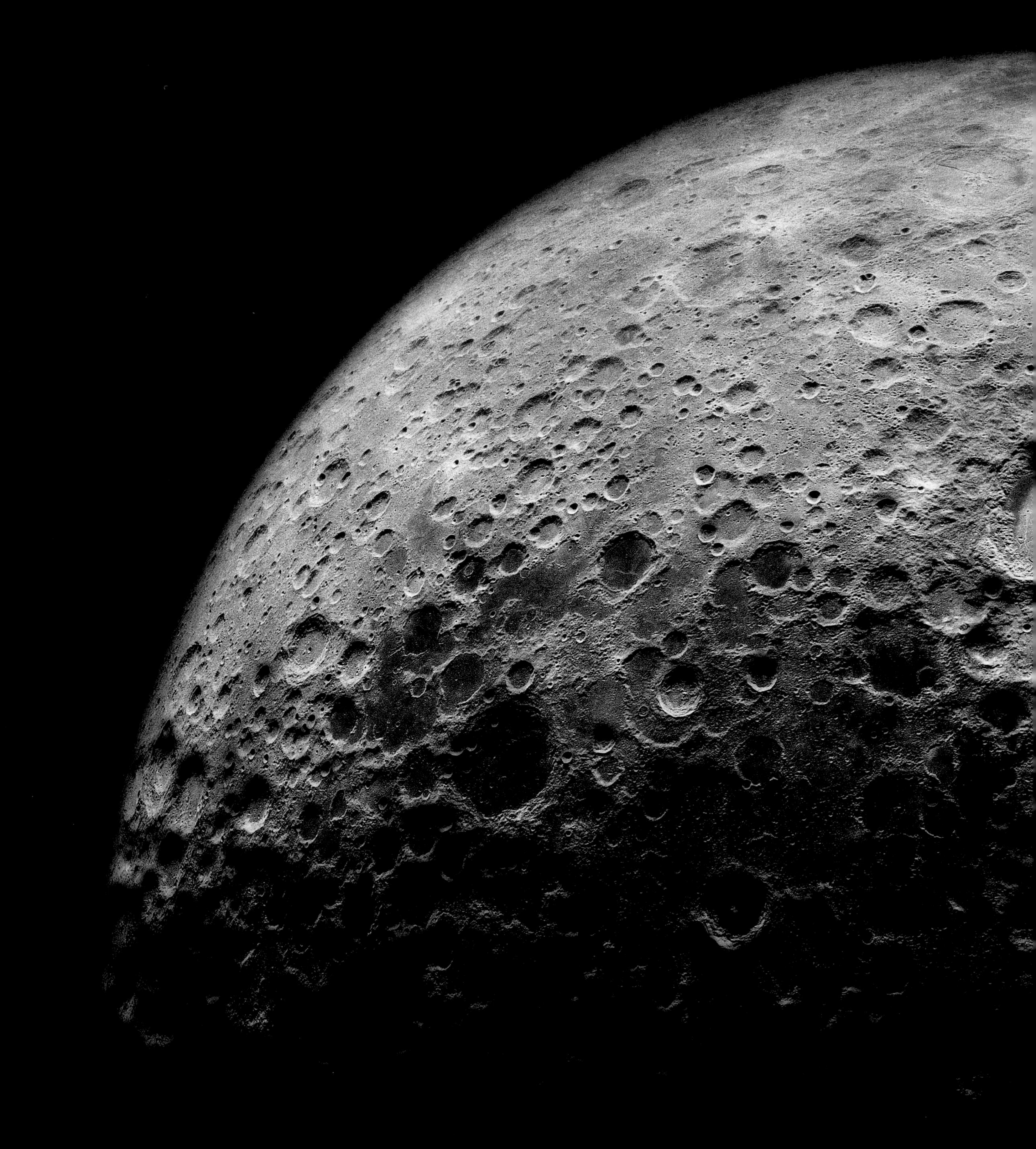

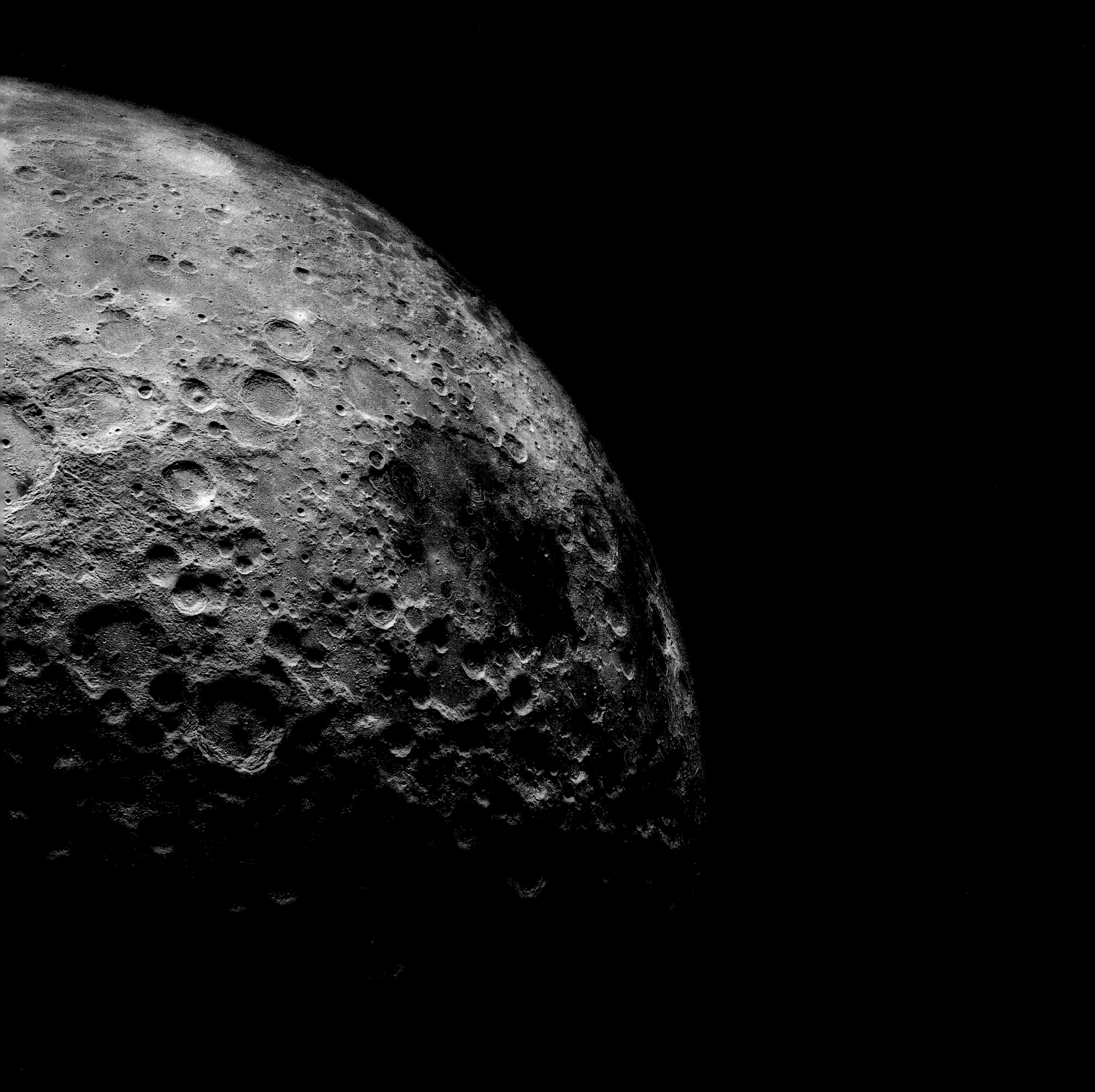

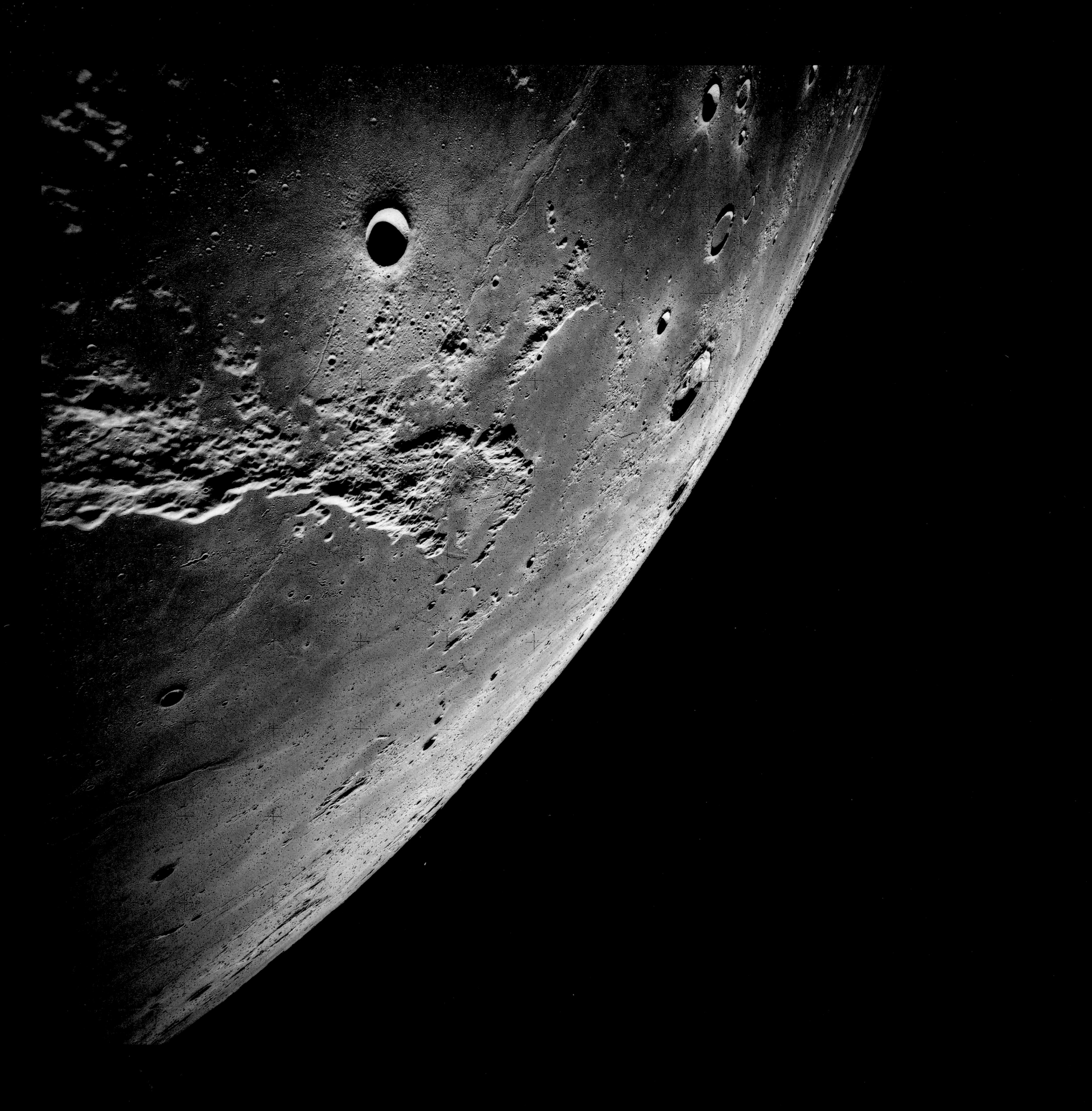

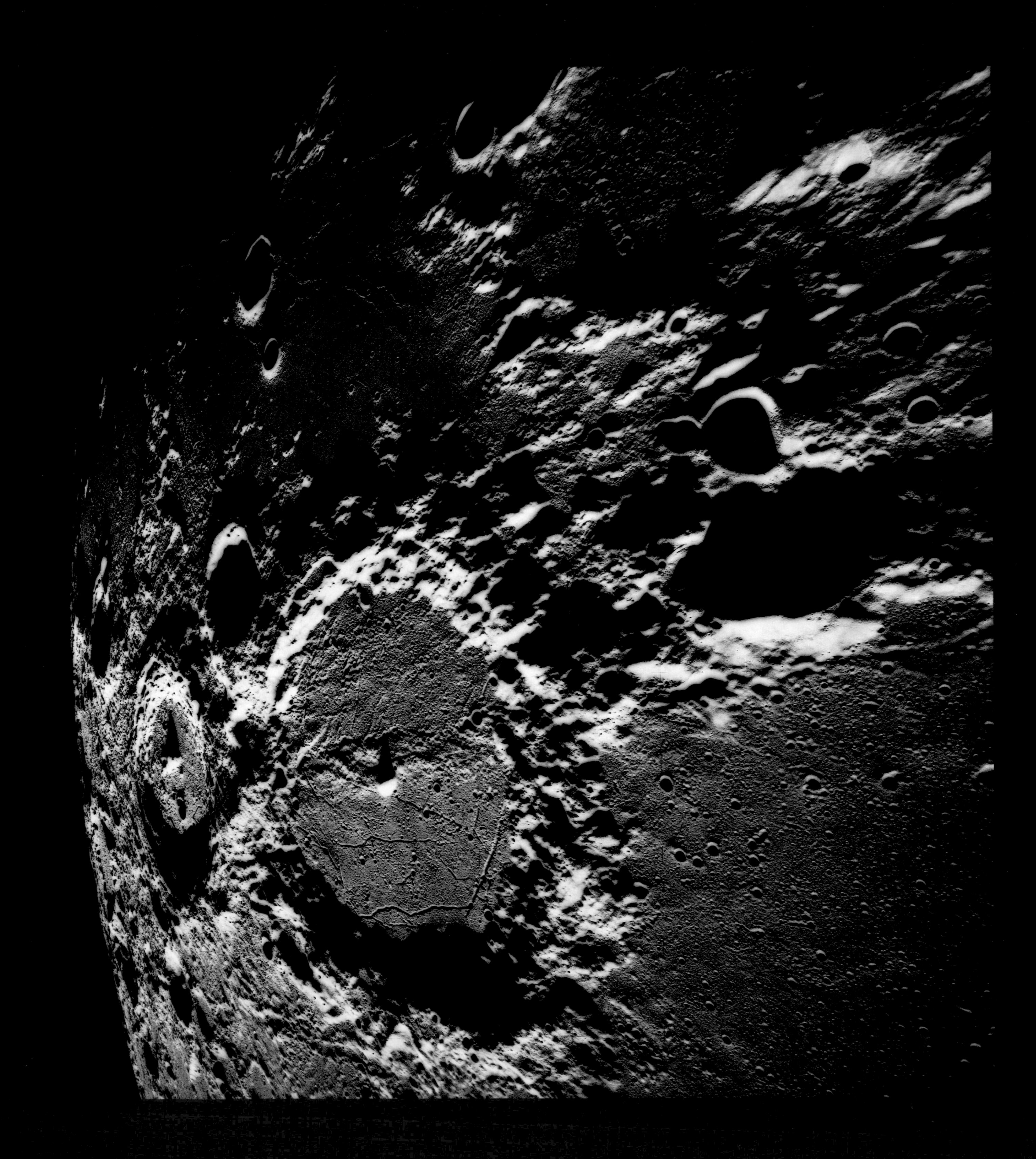

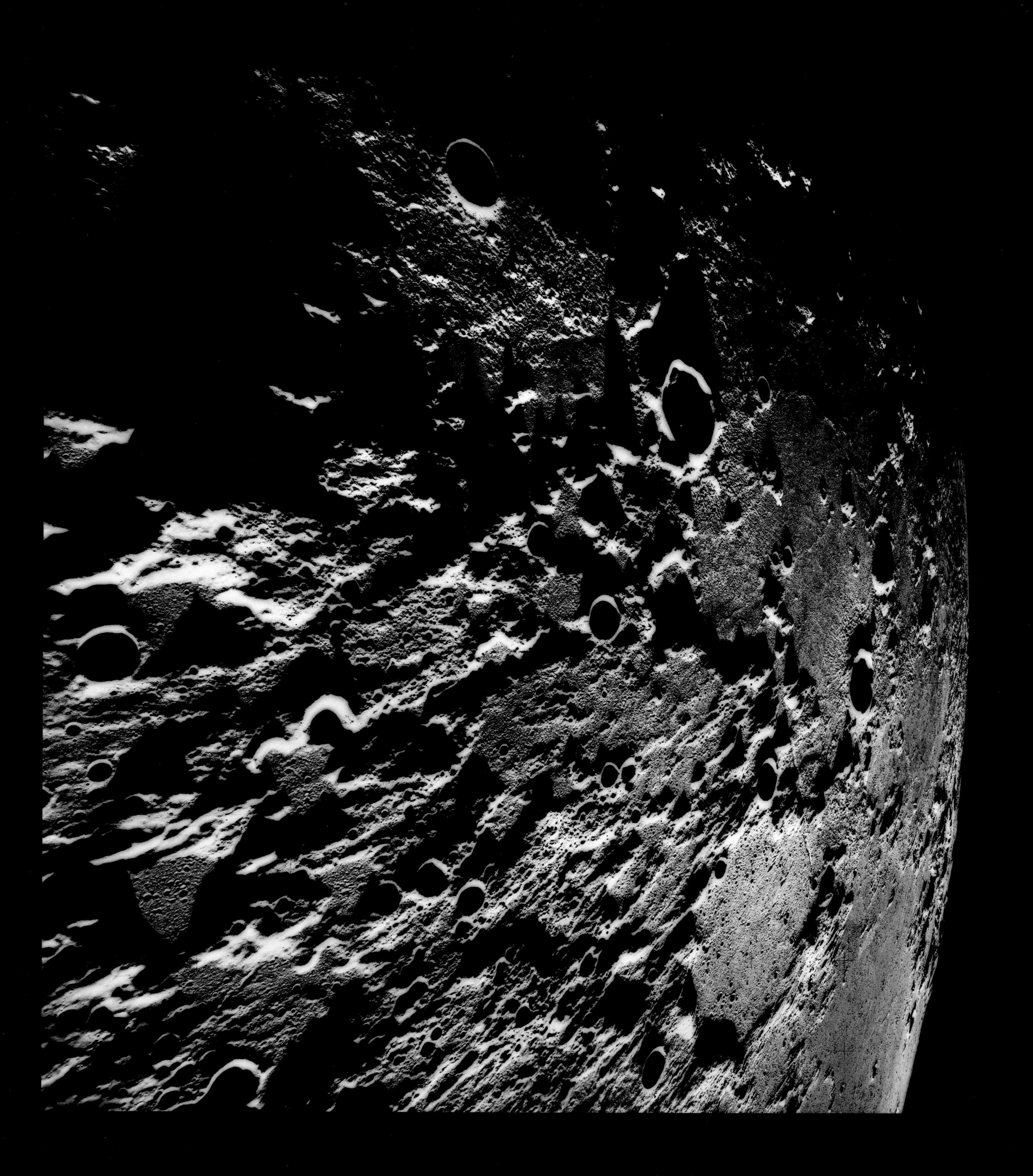

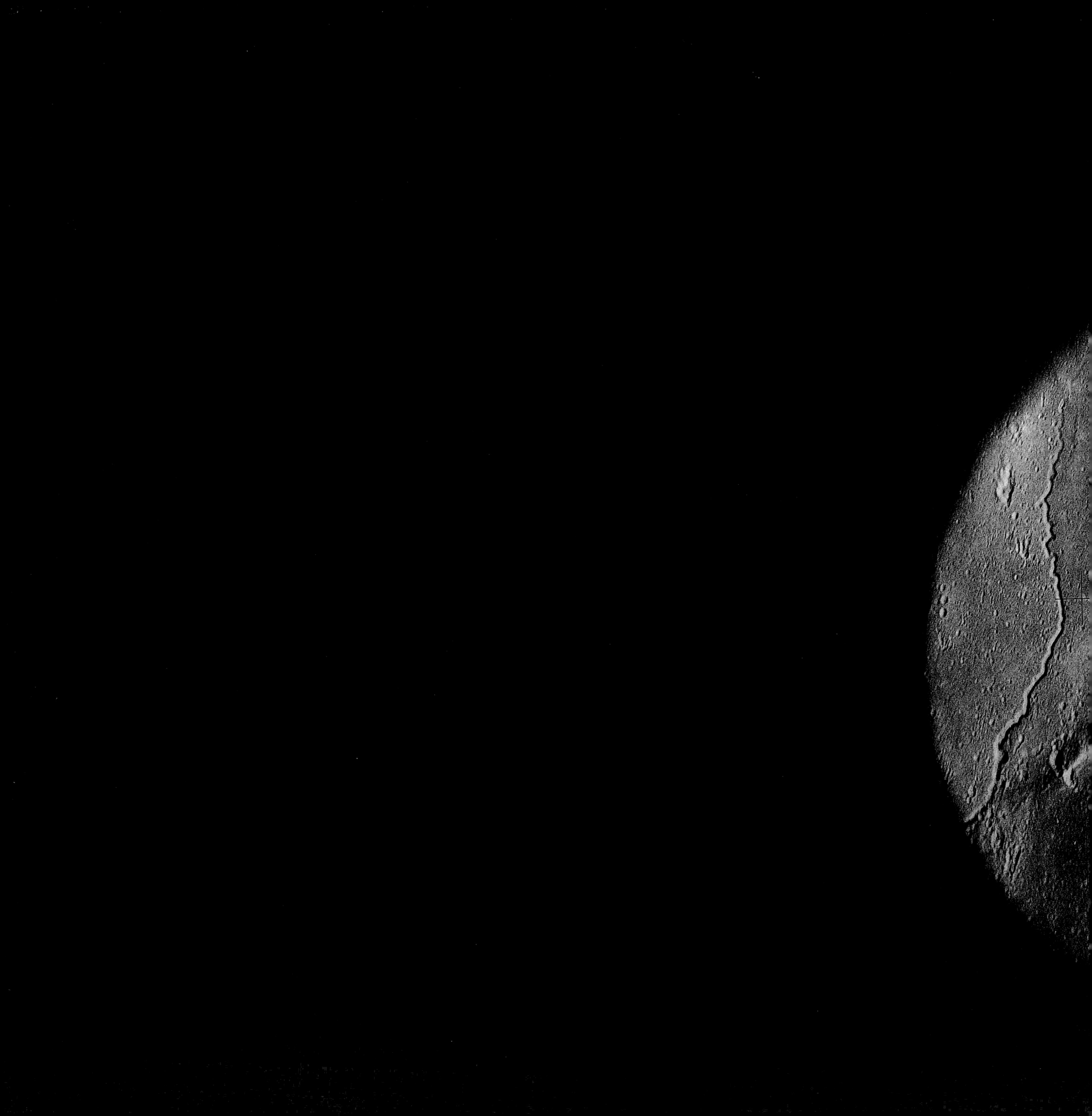

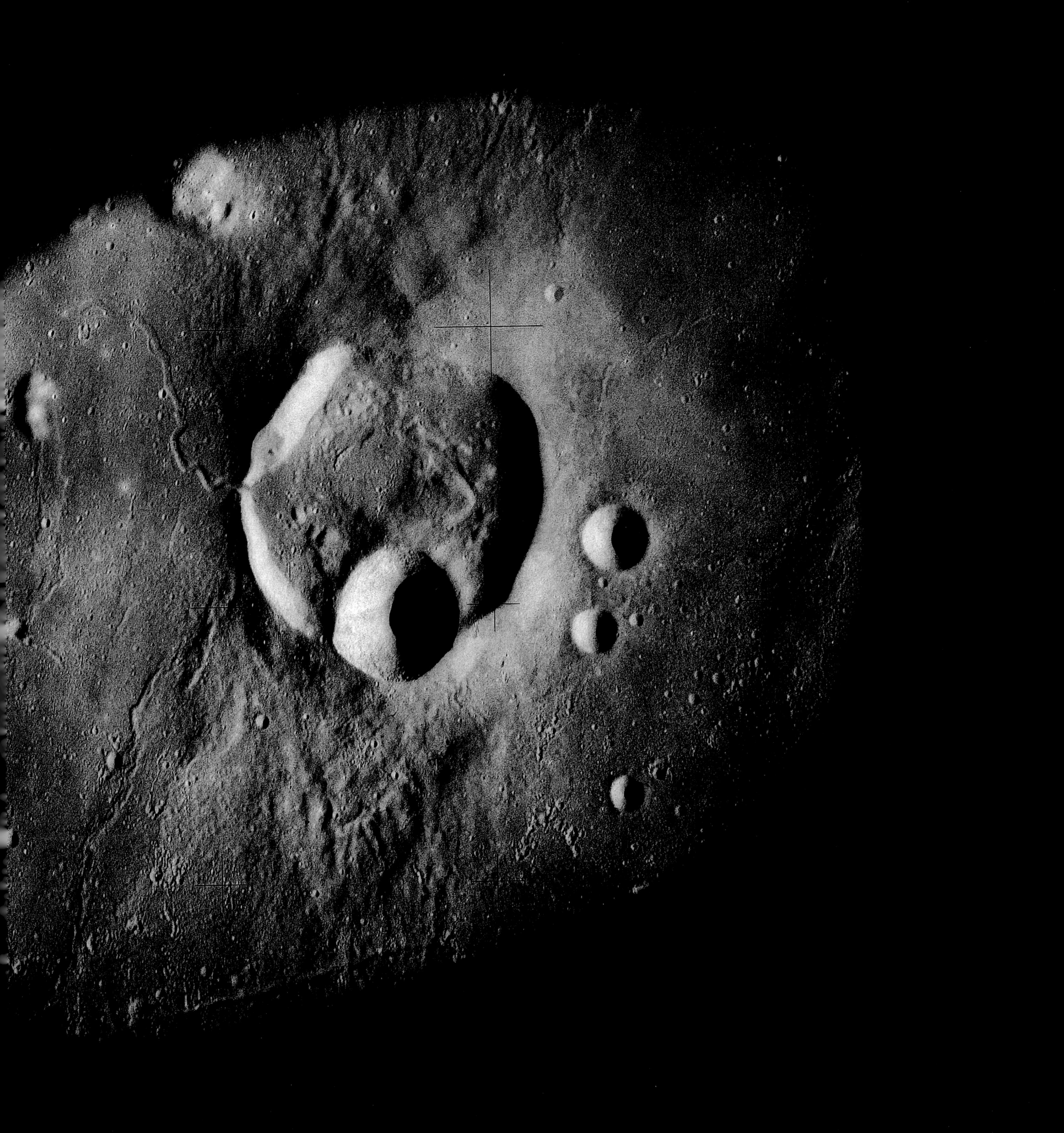

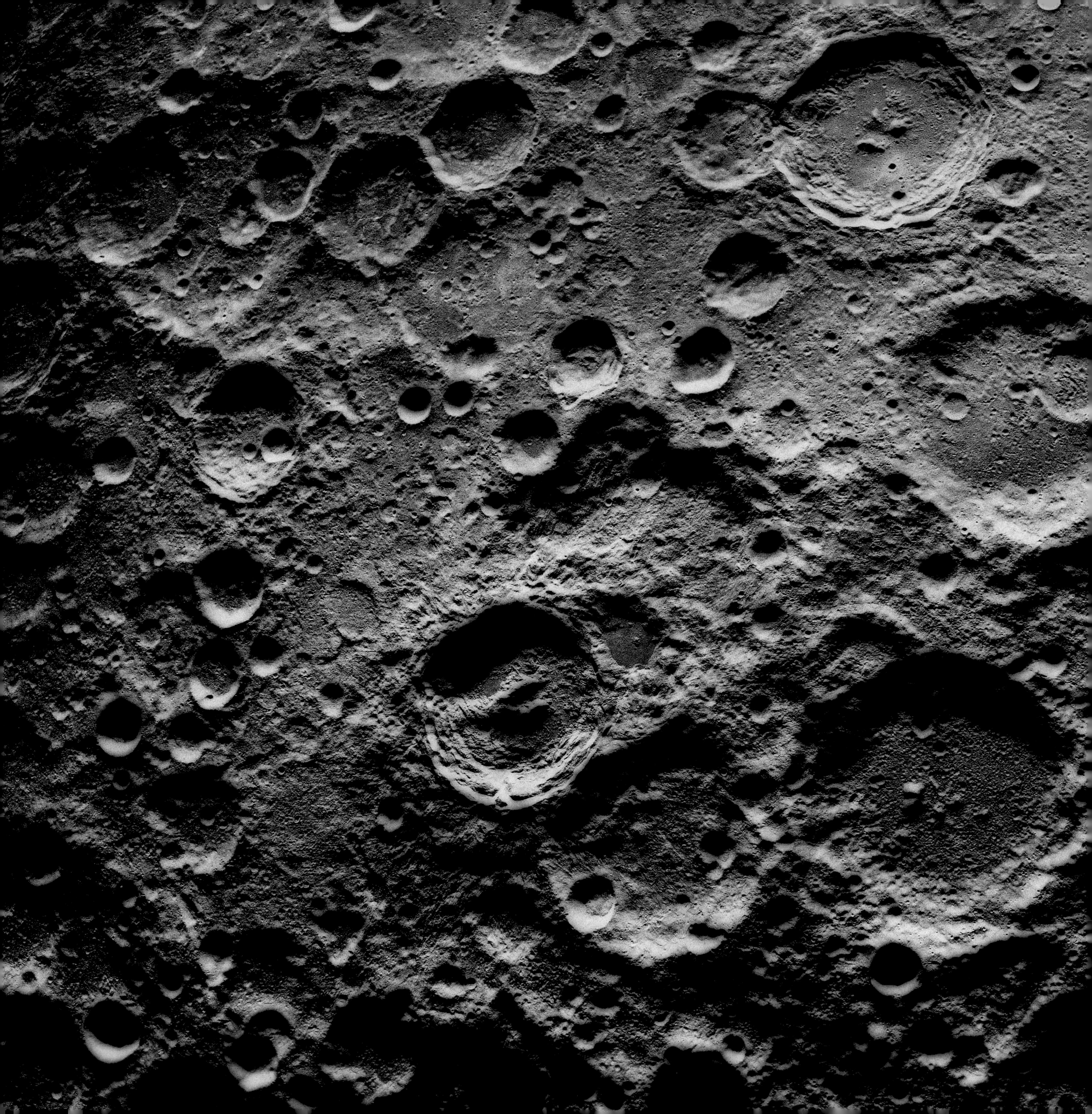

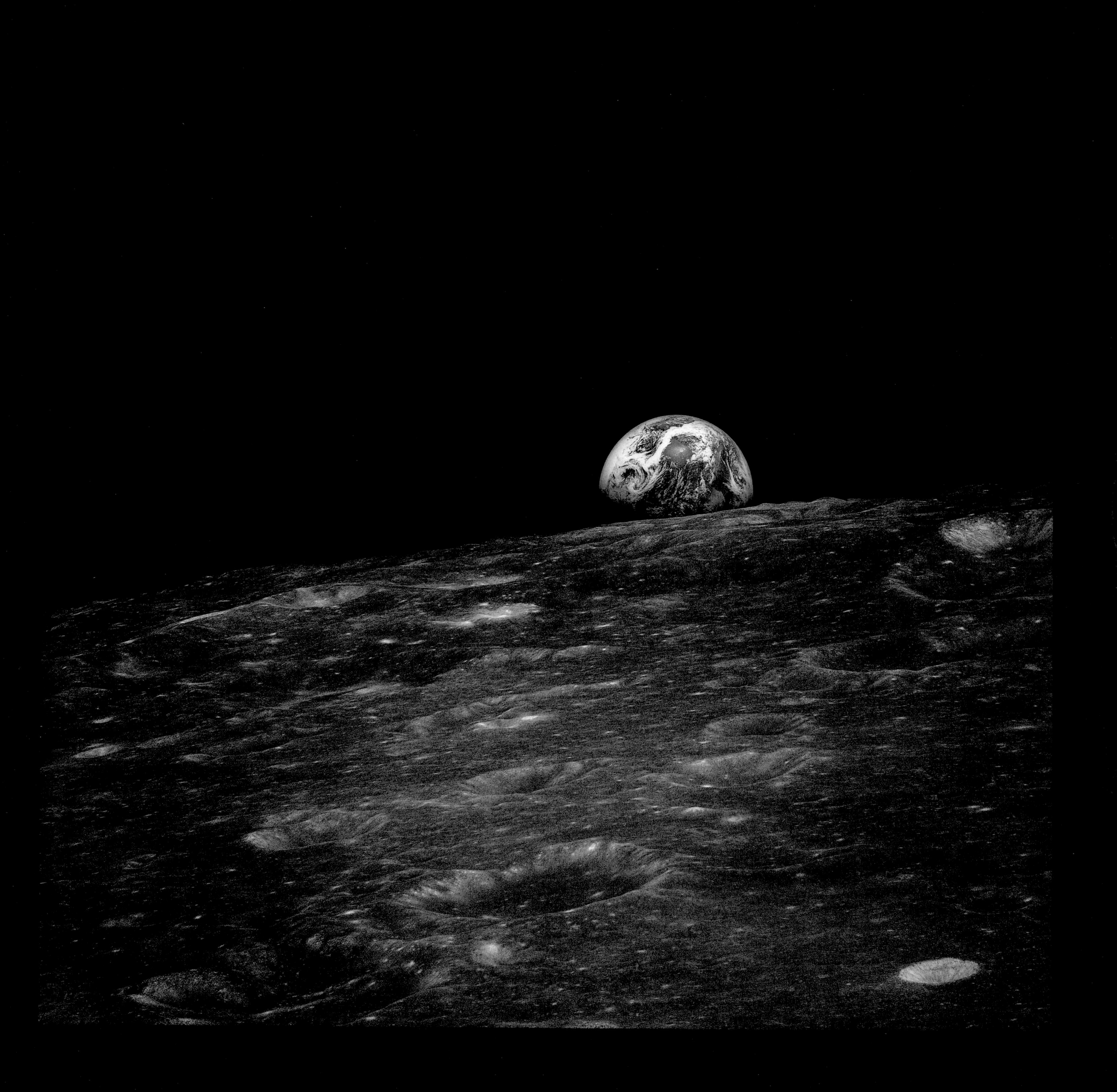

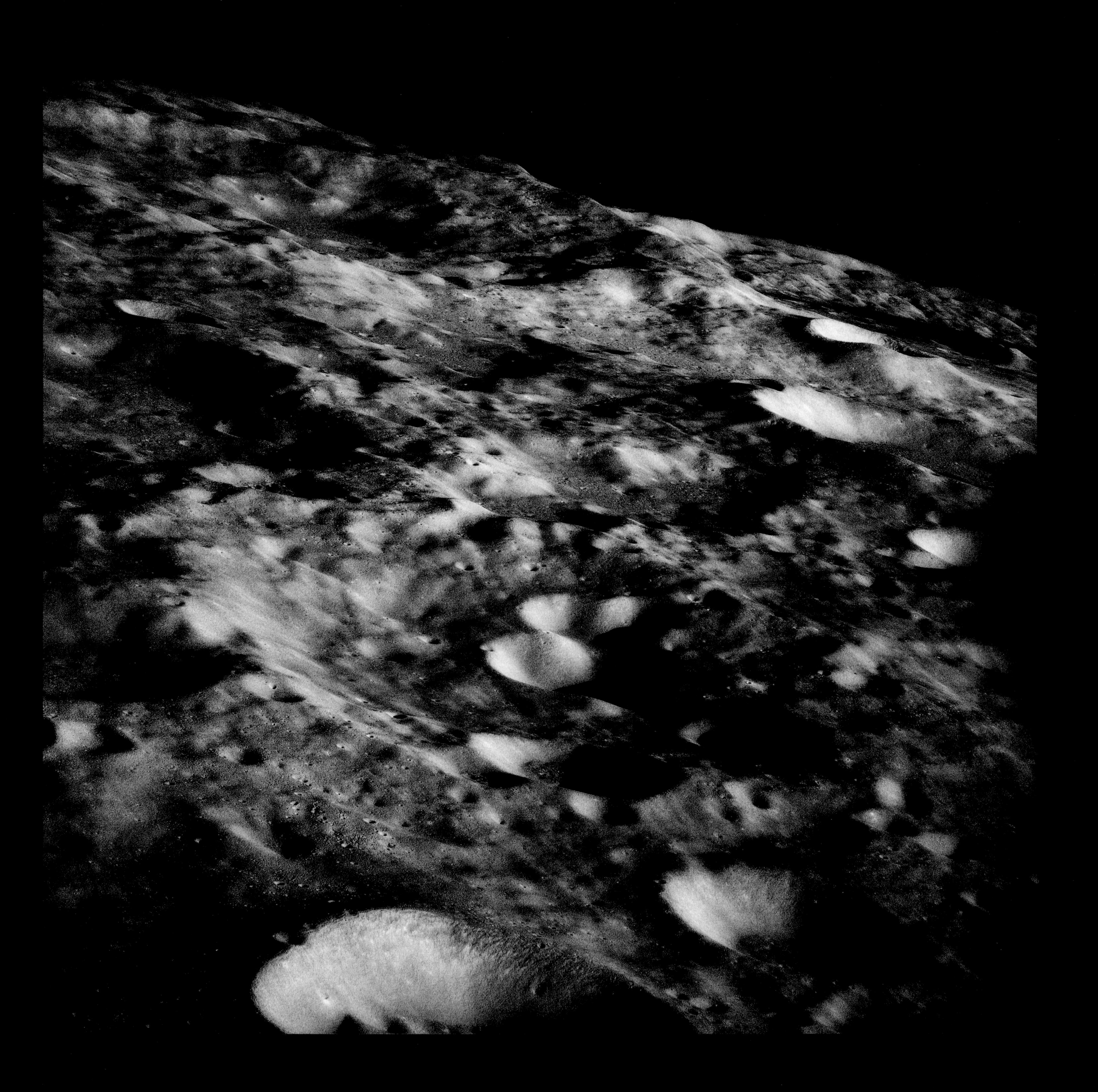

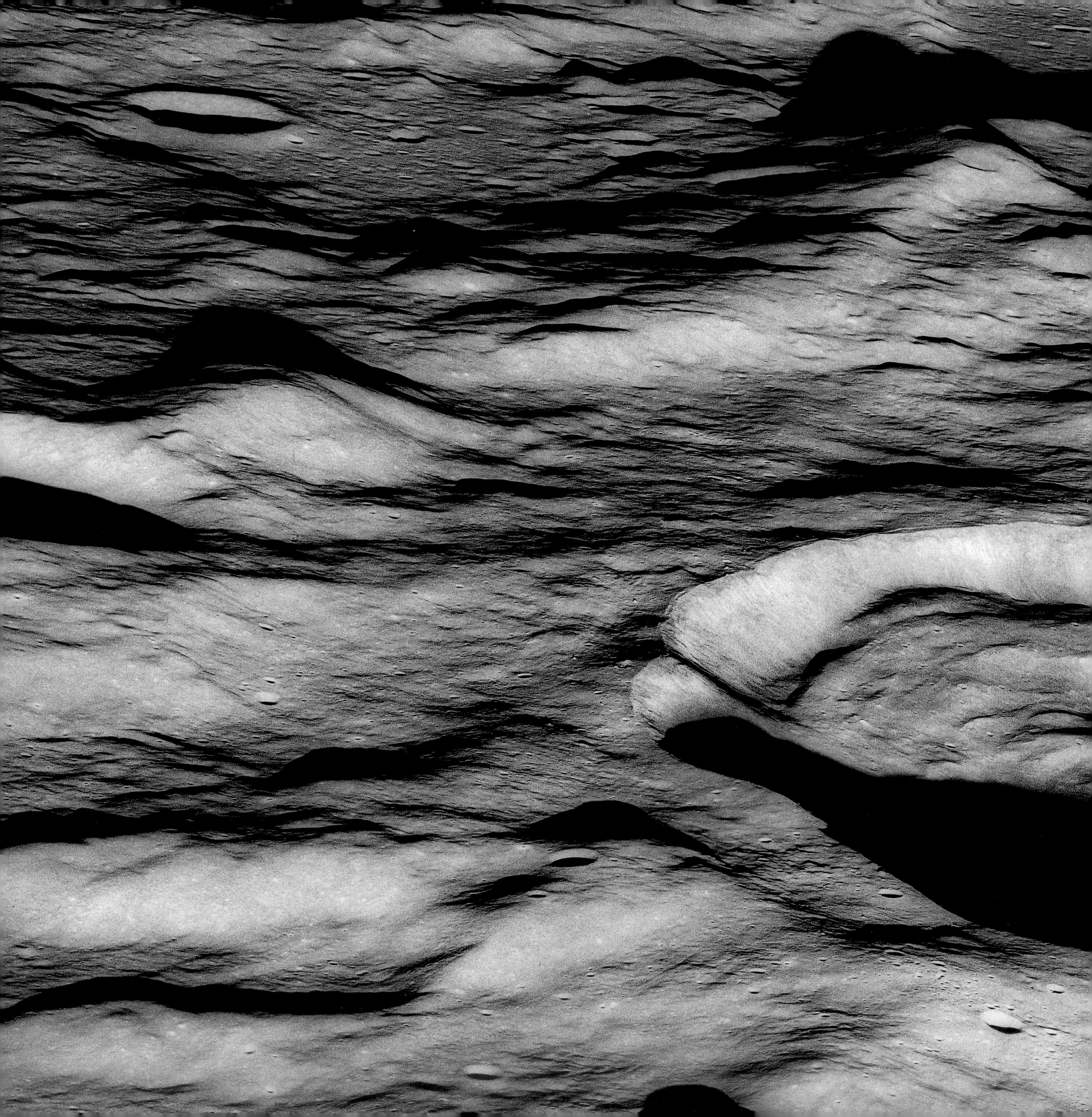

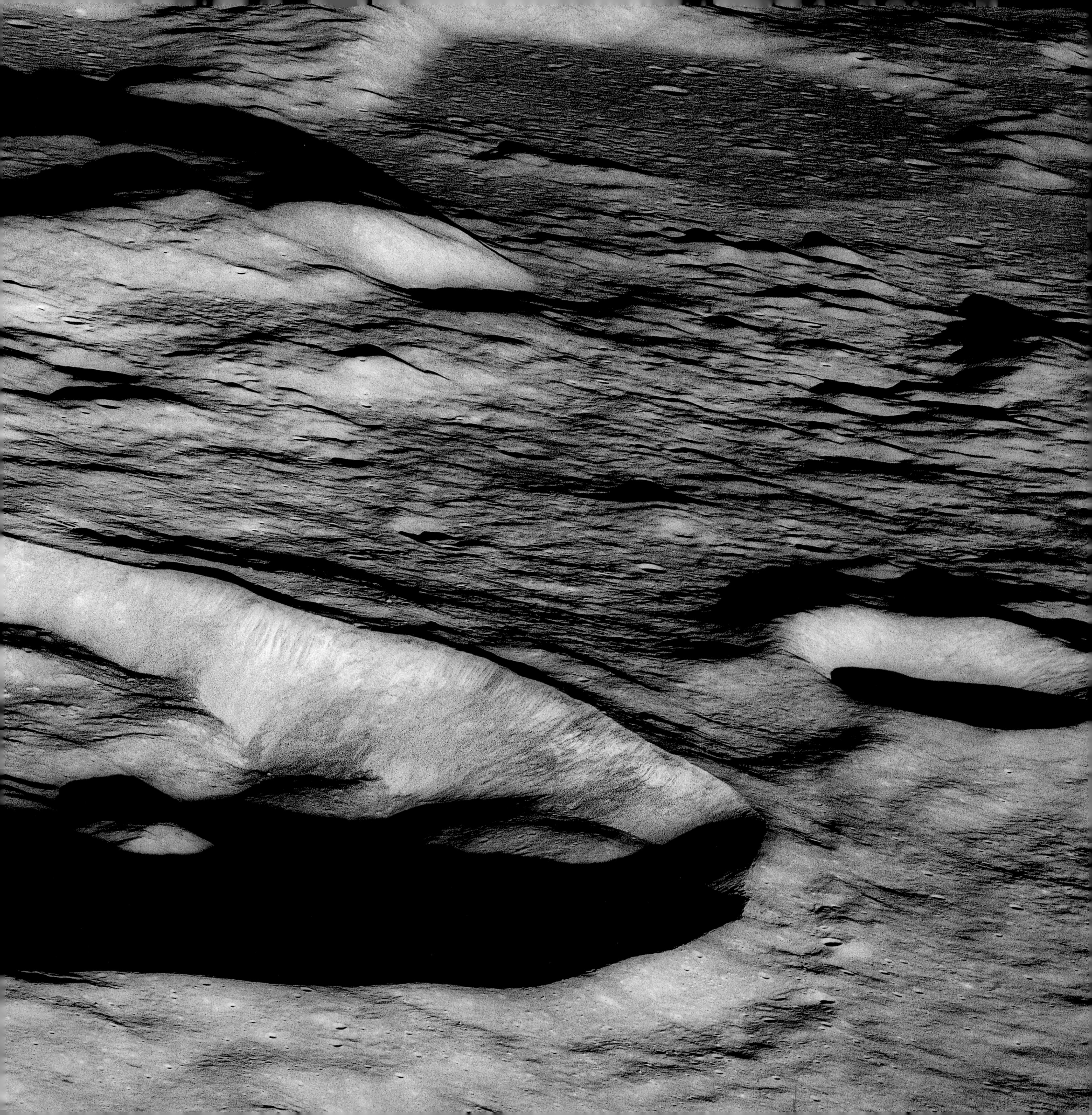

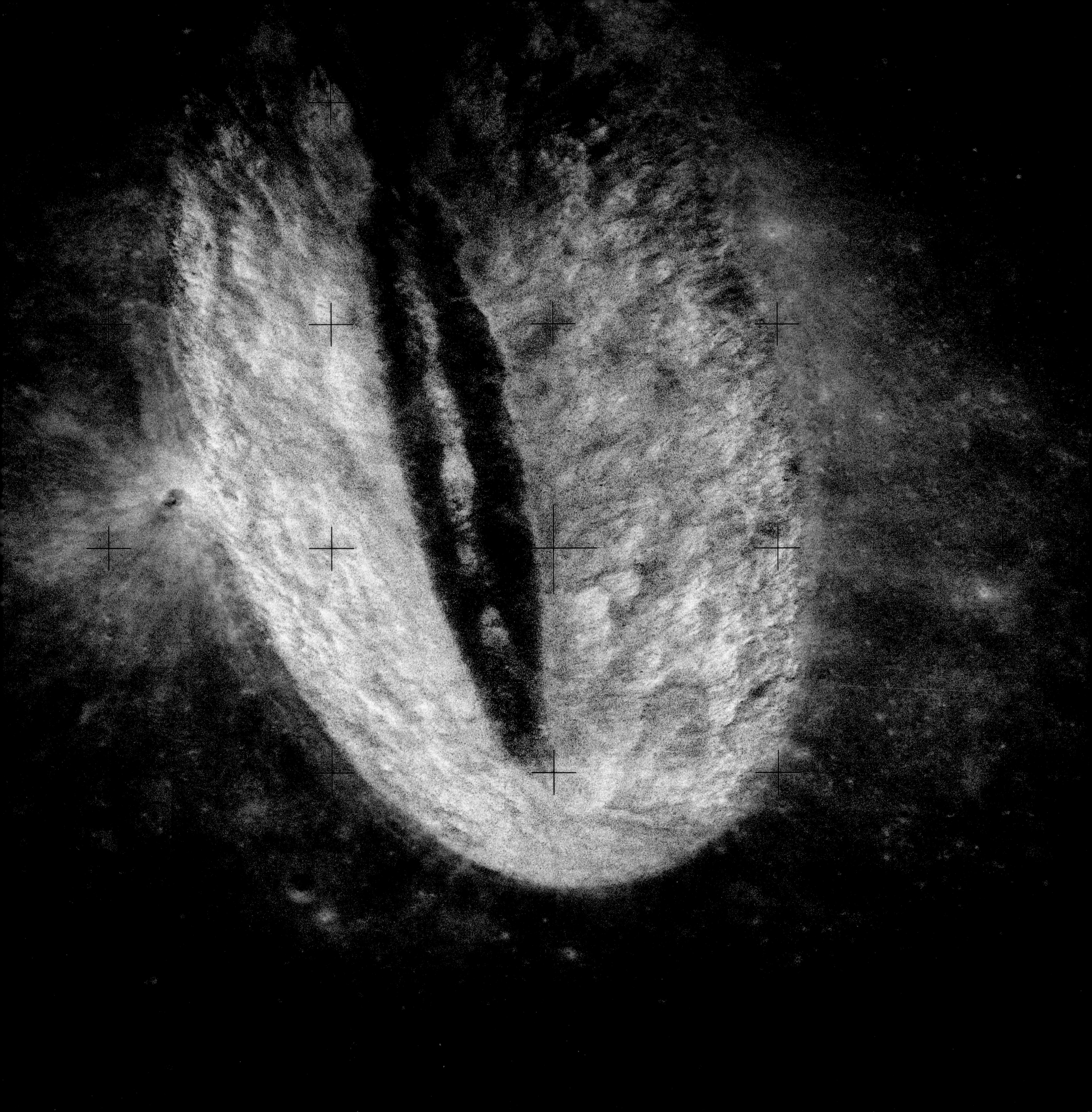

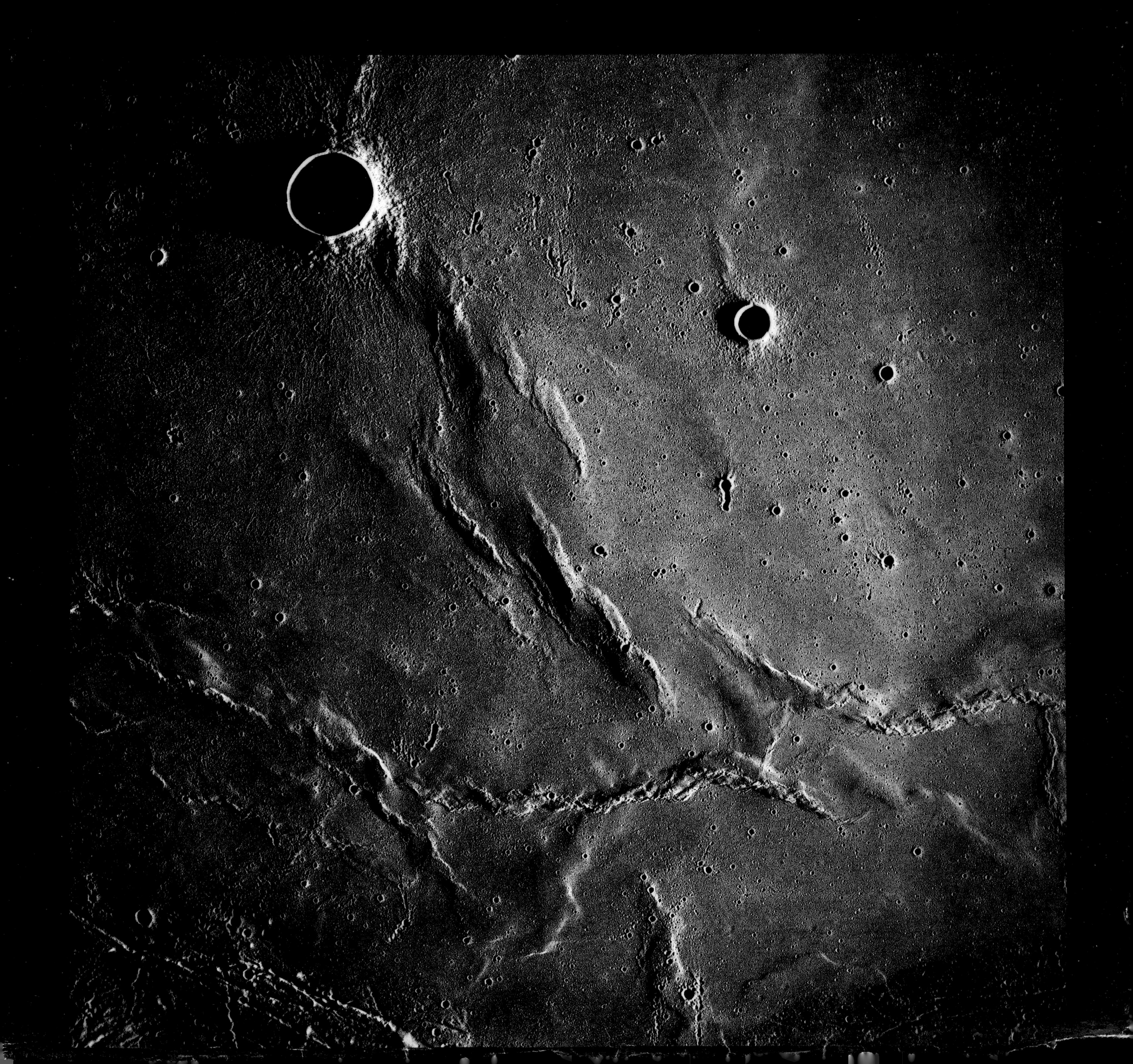

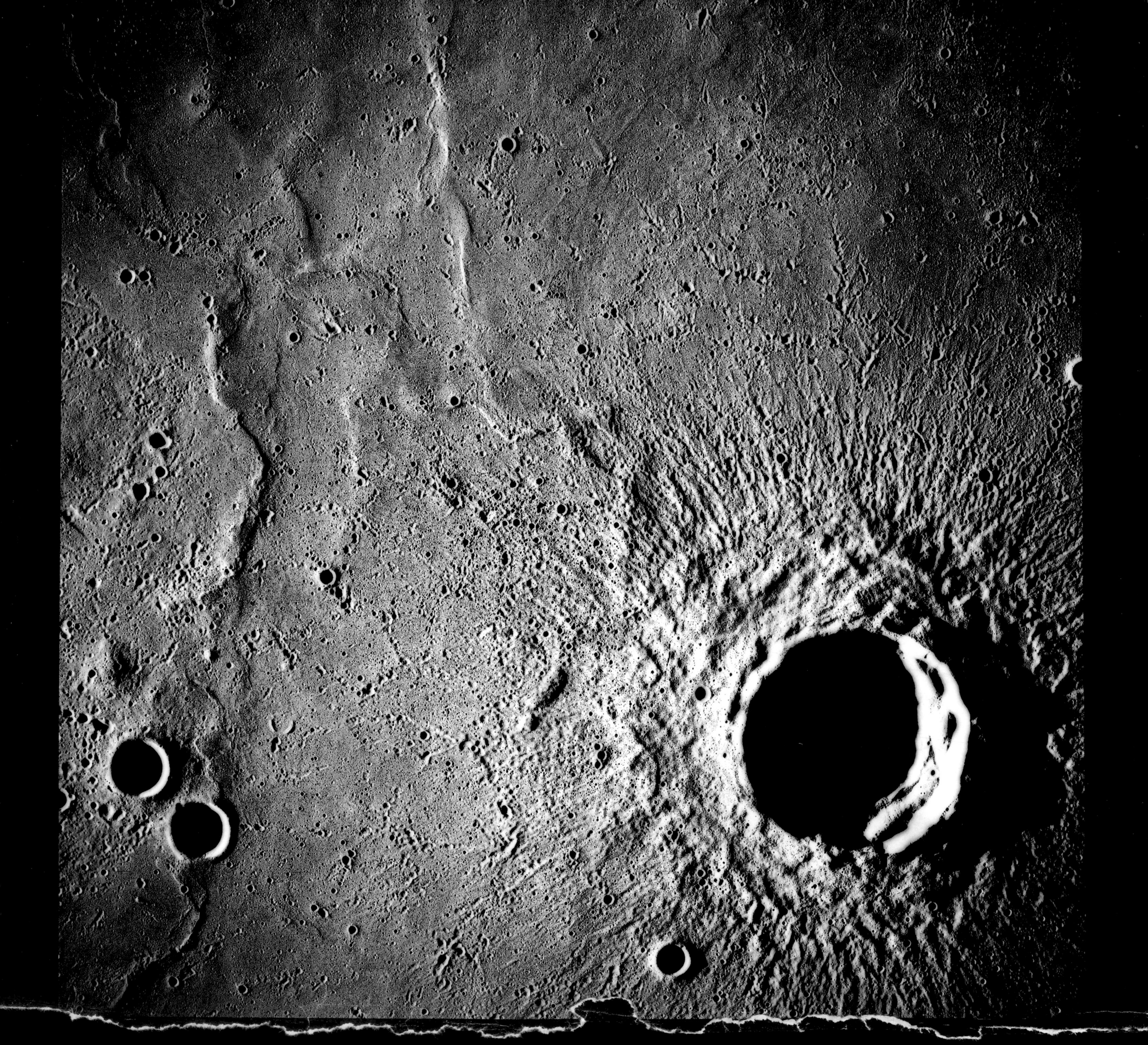

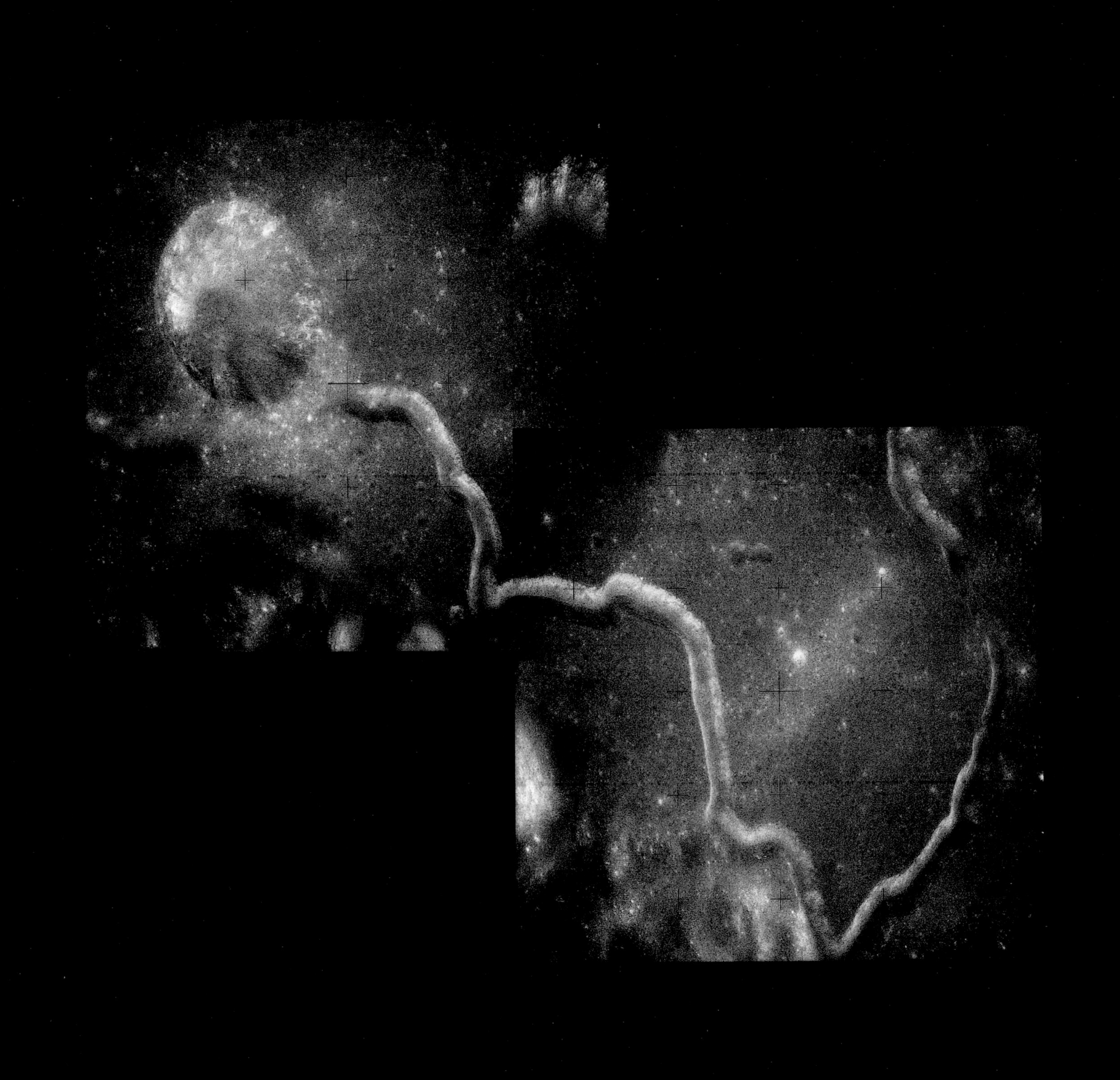

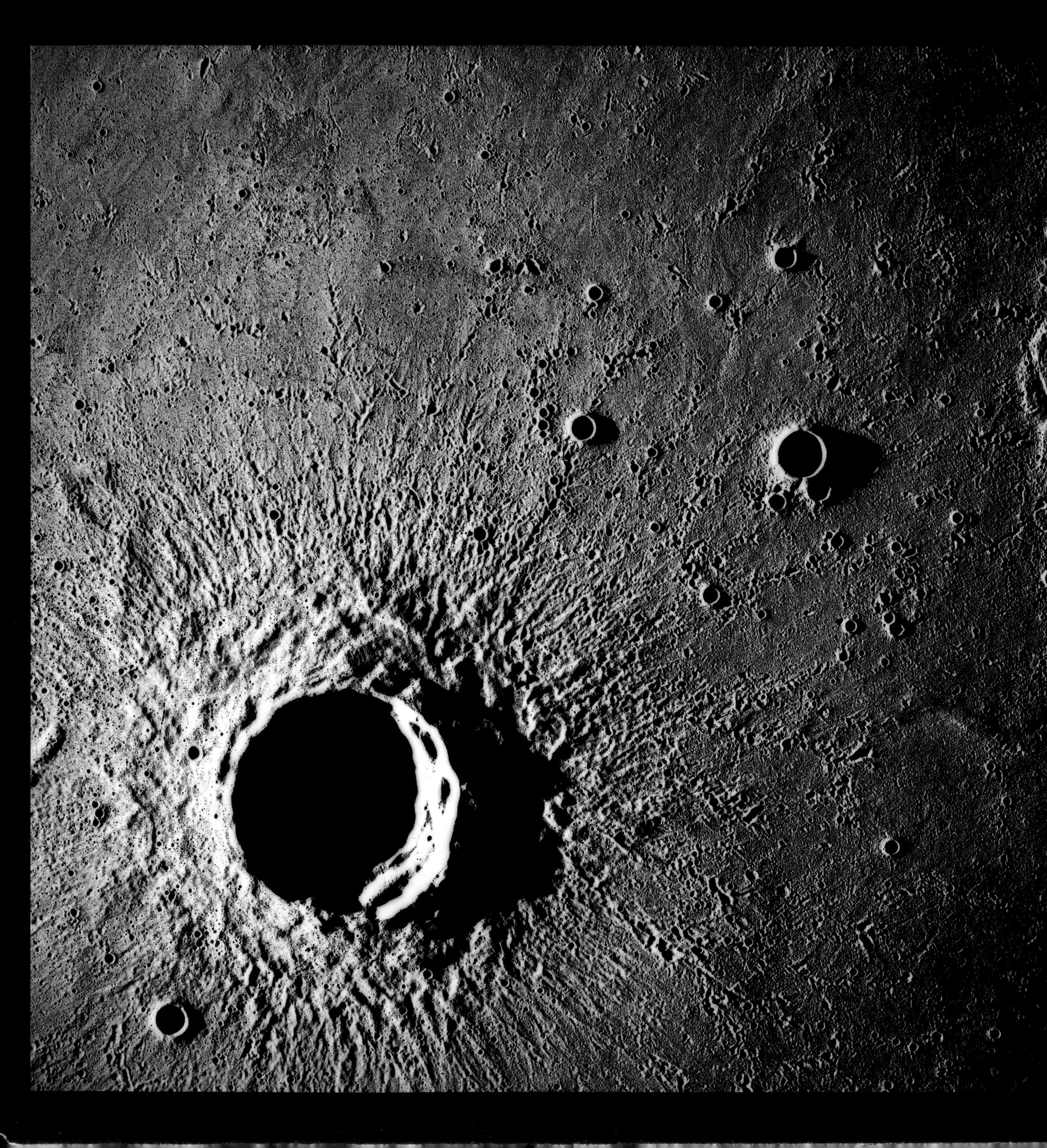

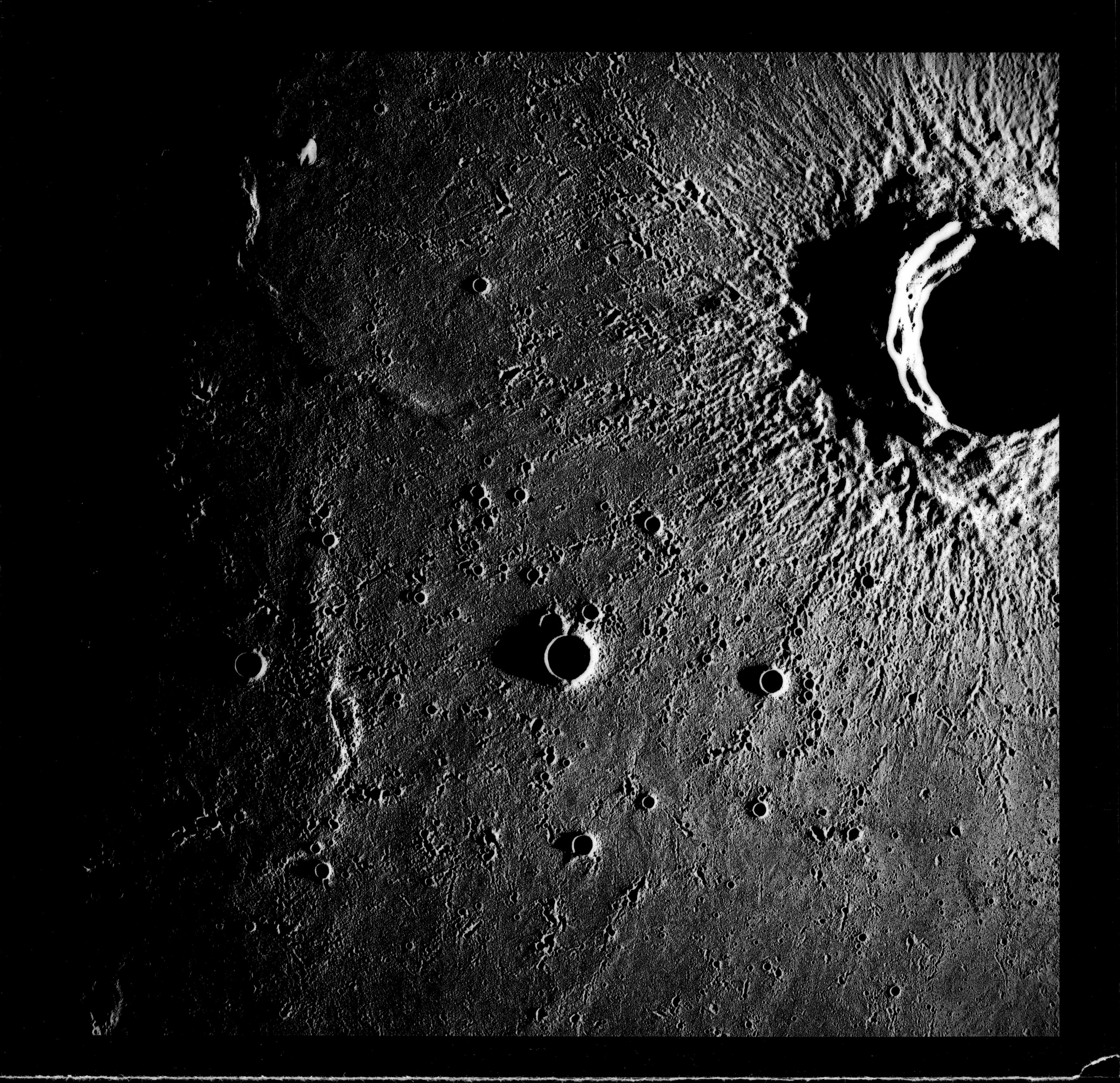

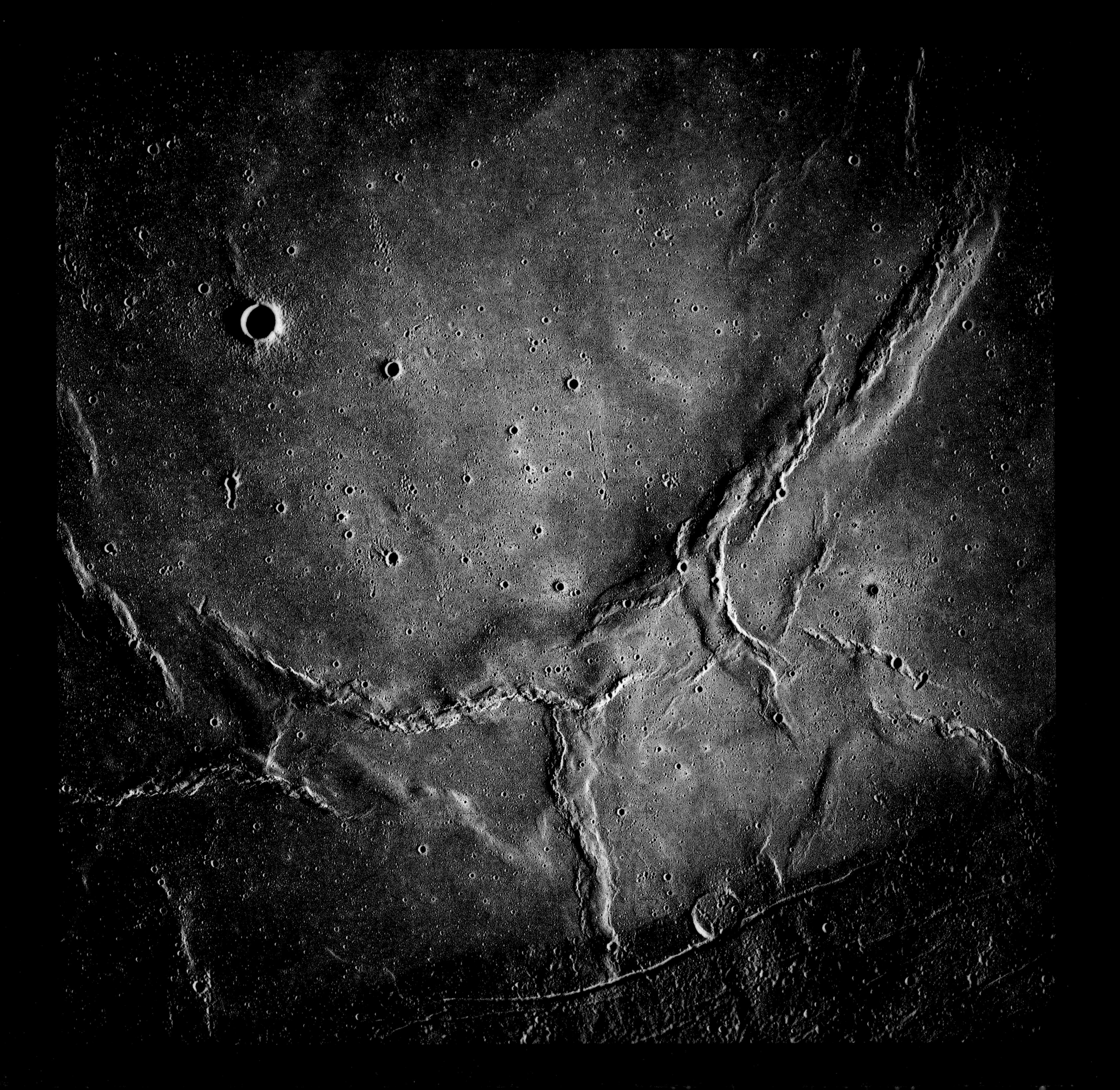

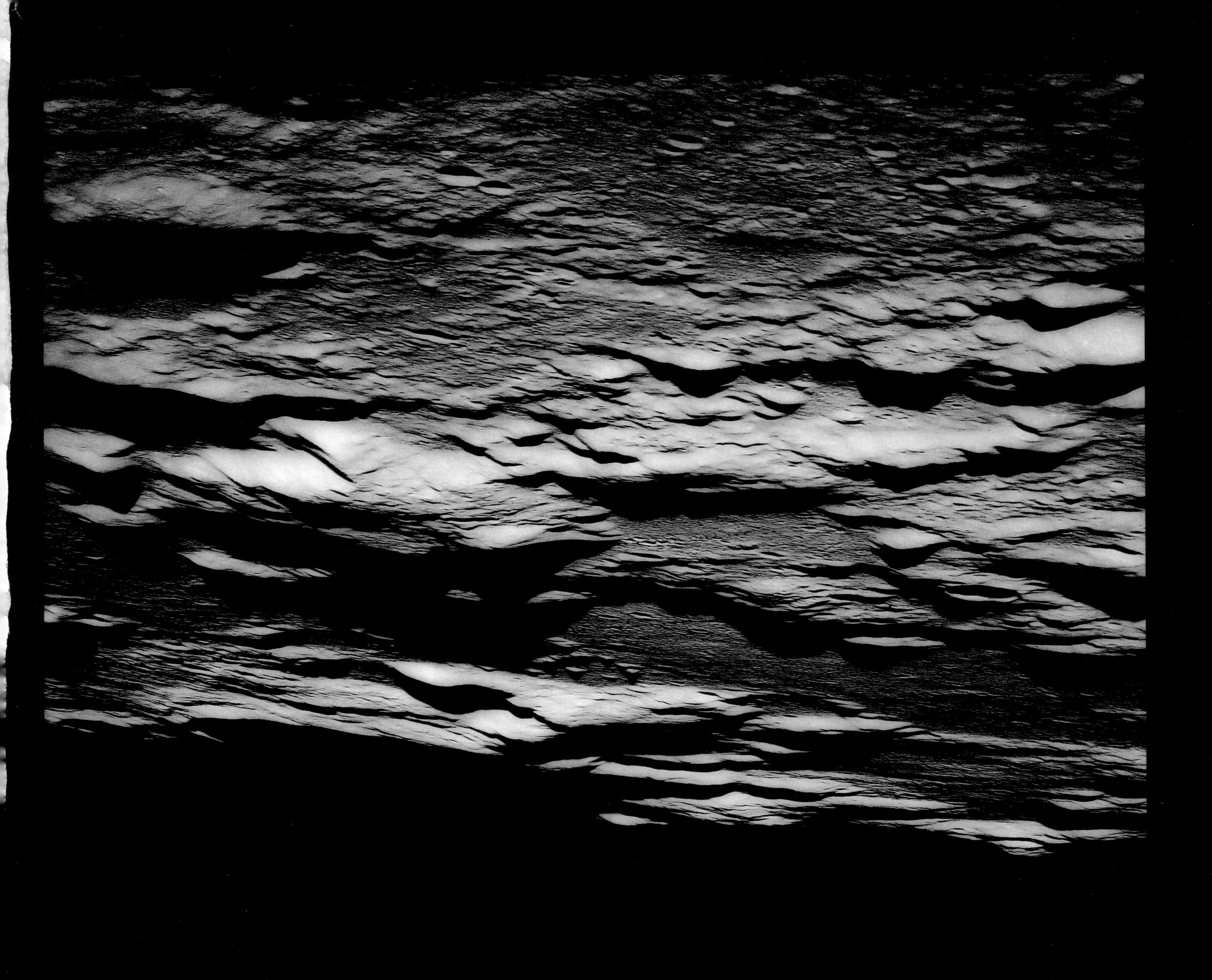

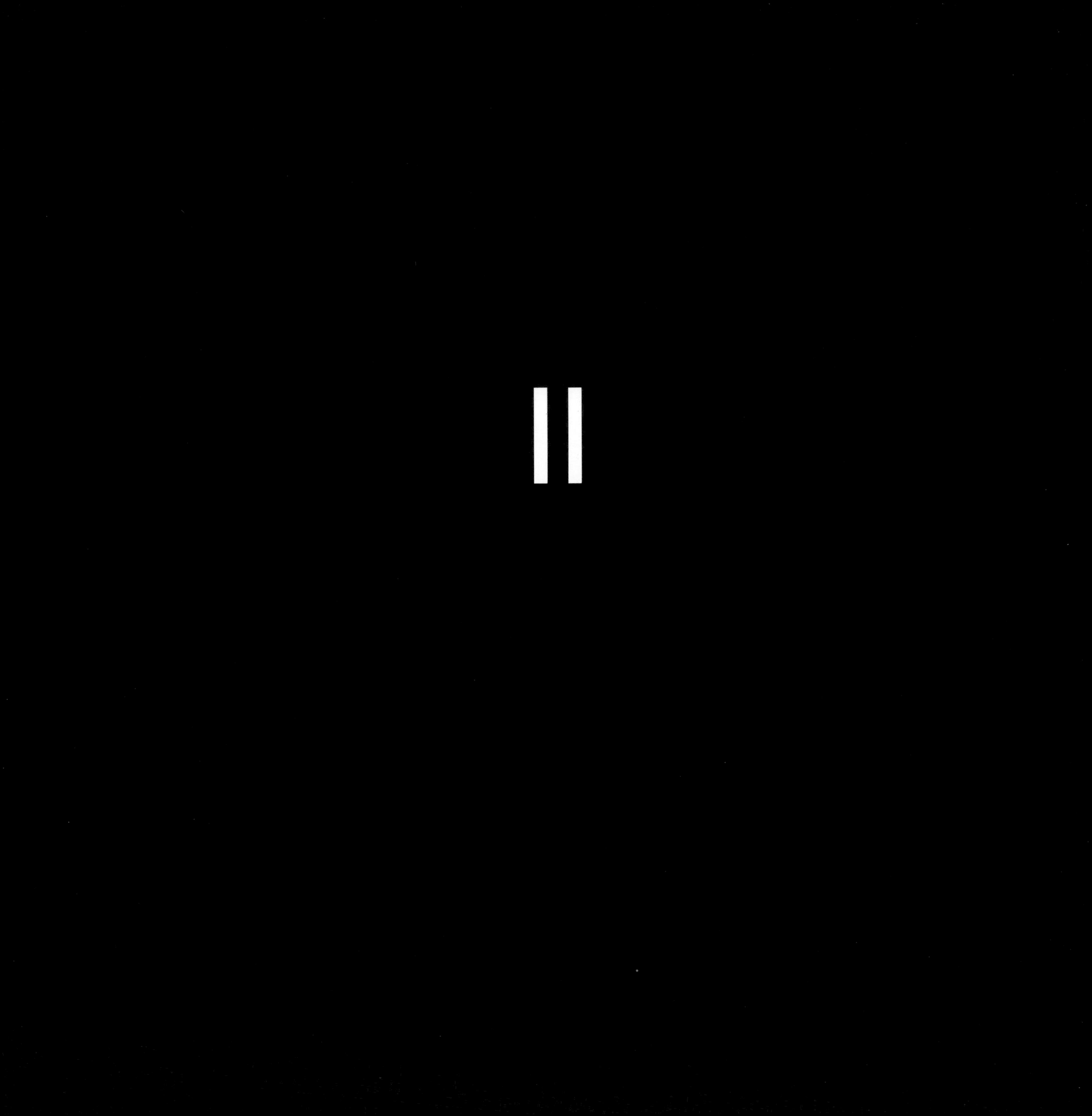
II

NASA

UNITED
STATES

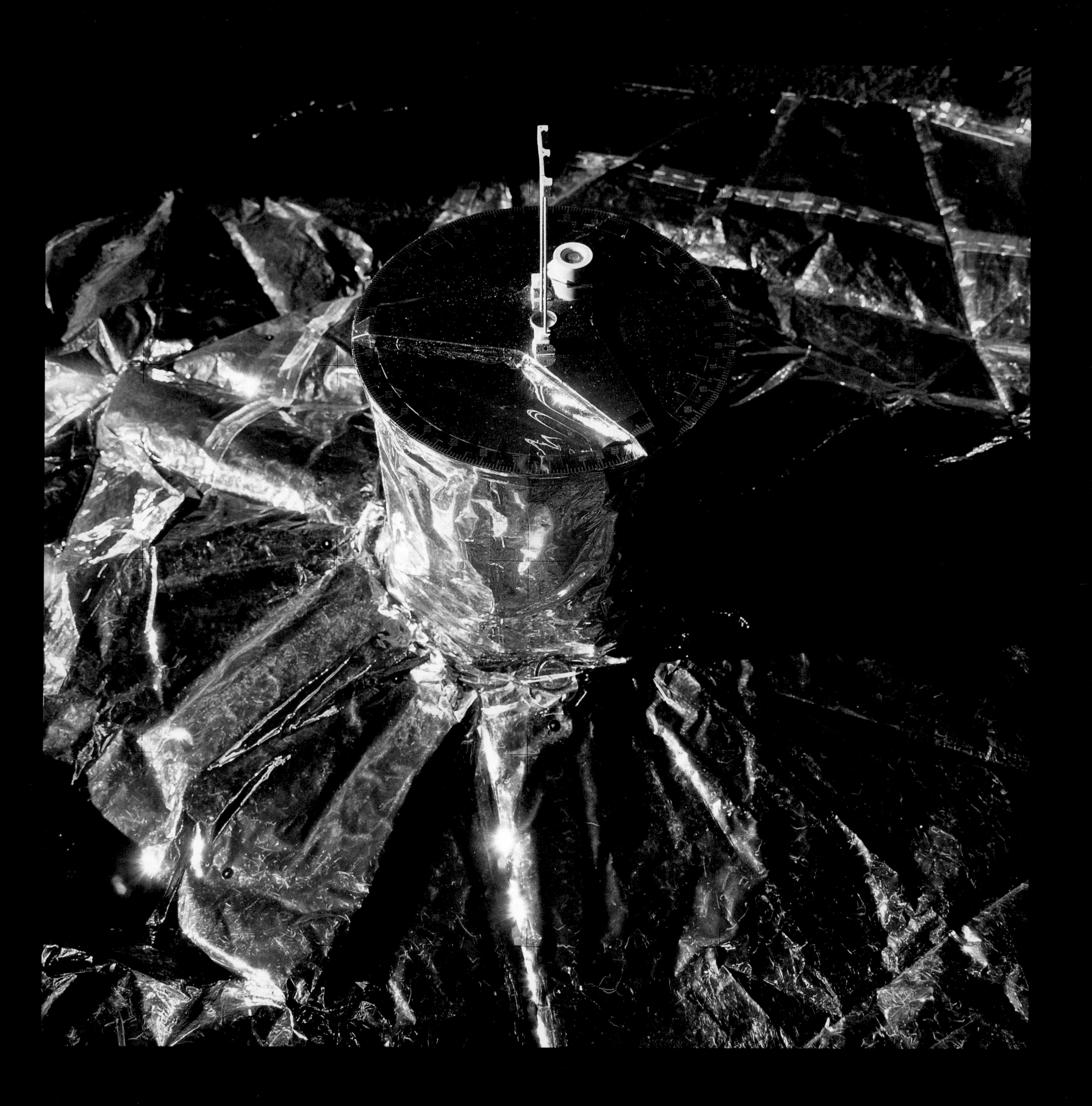

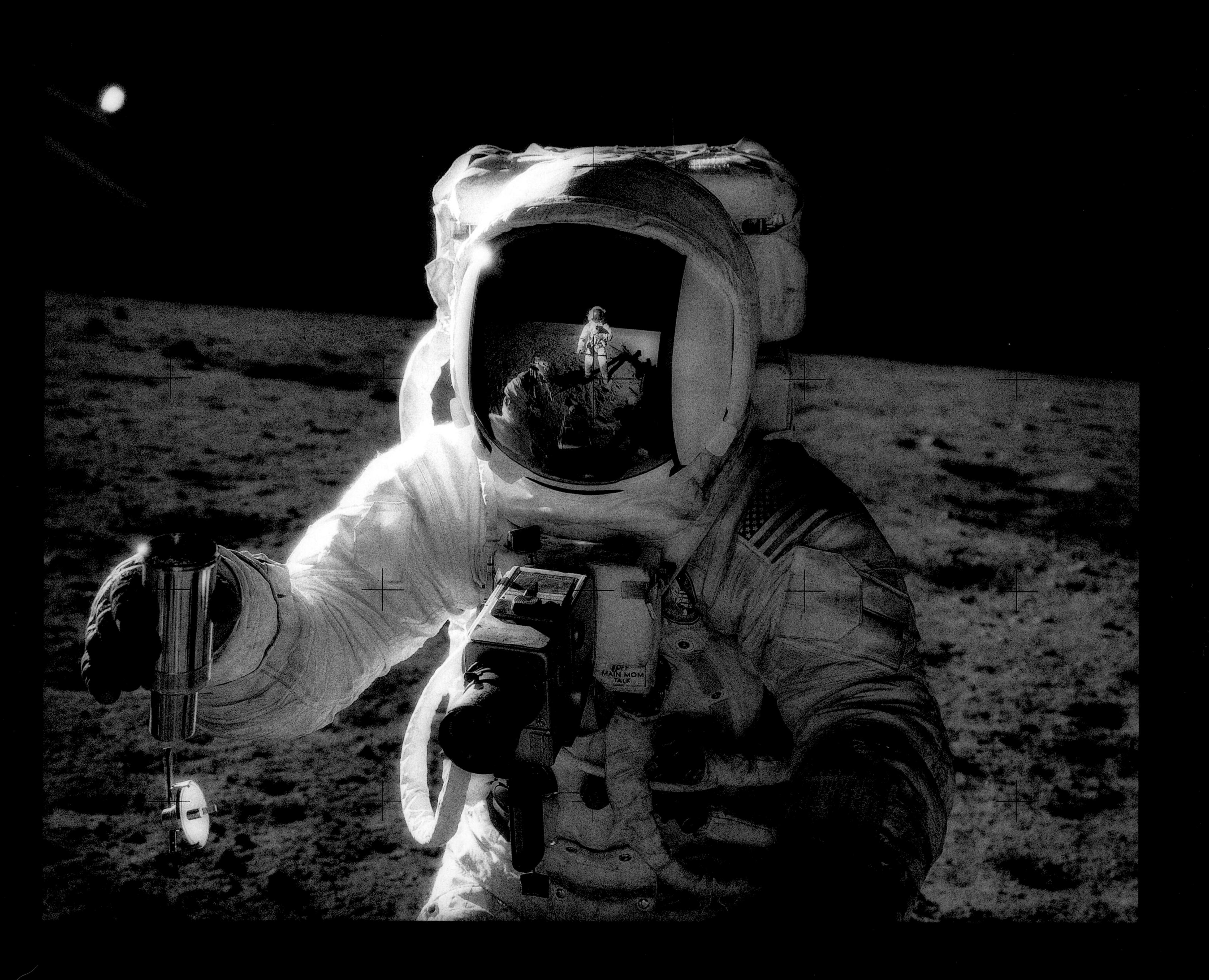
MAIN MOM
TALK

NASA

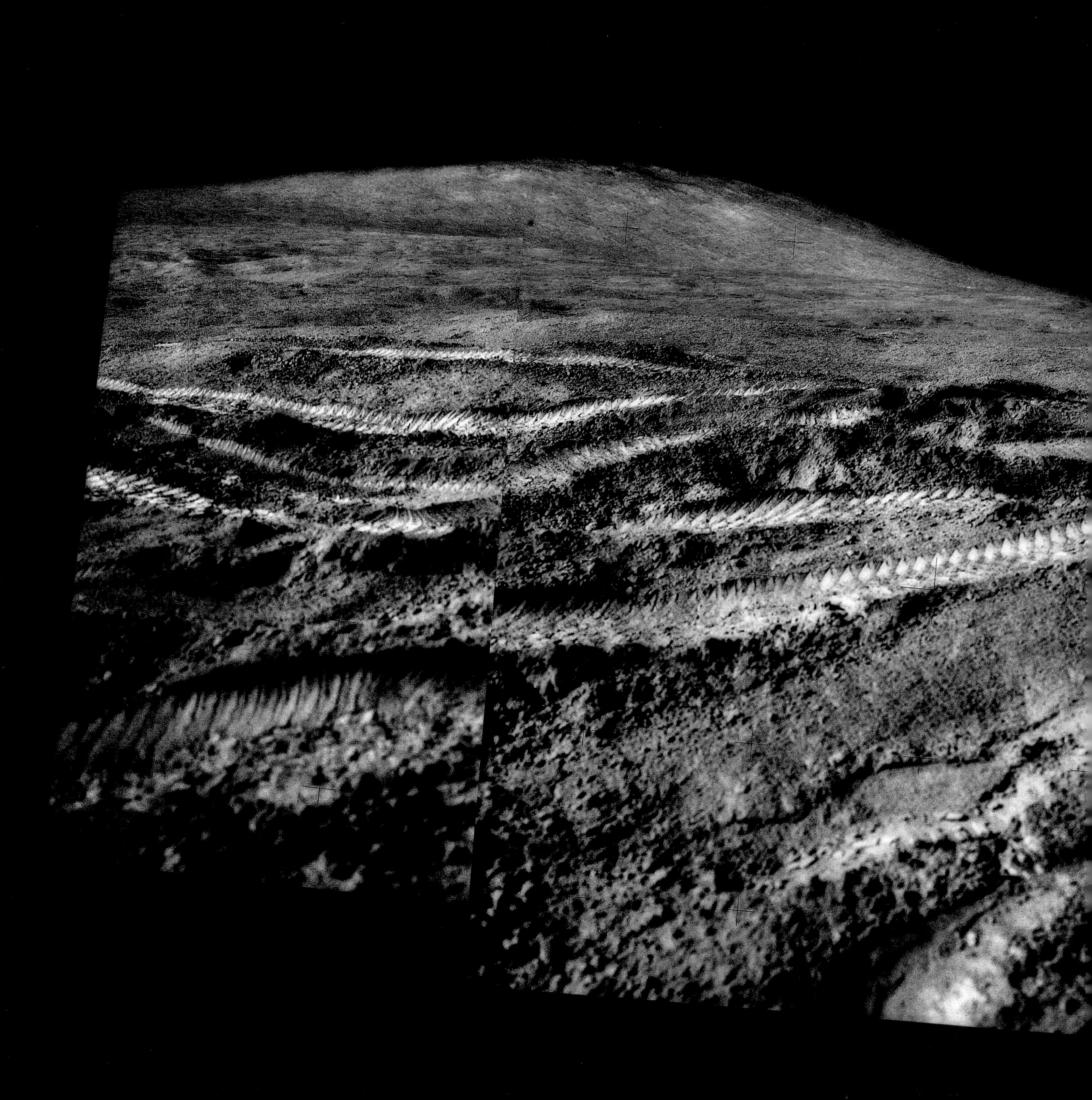

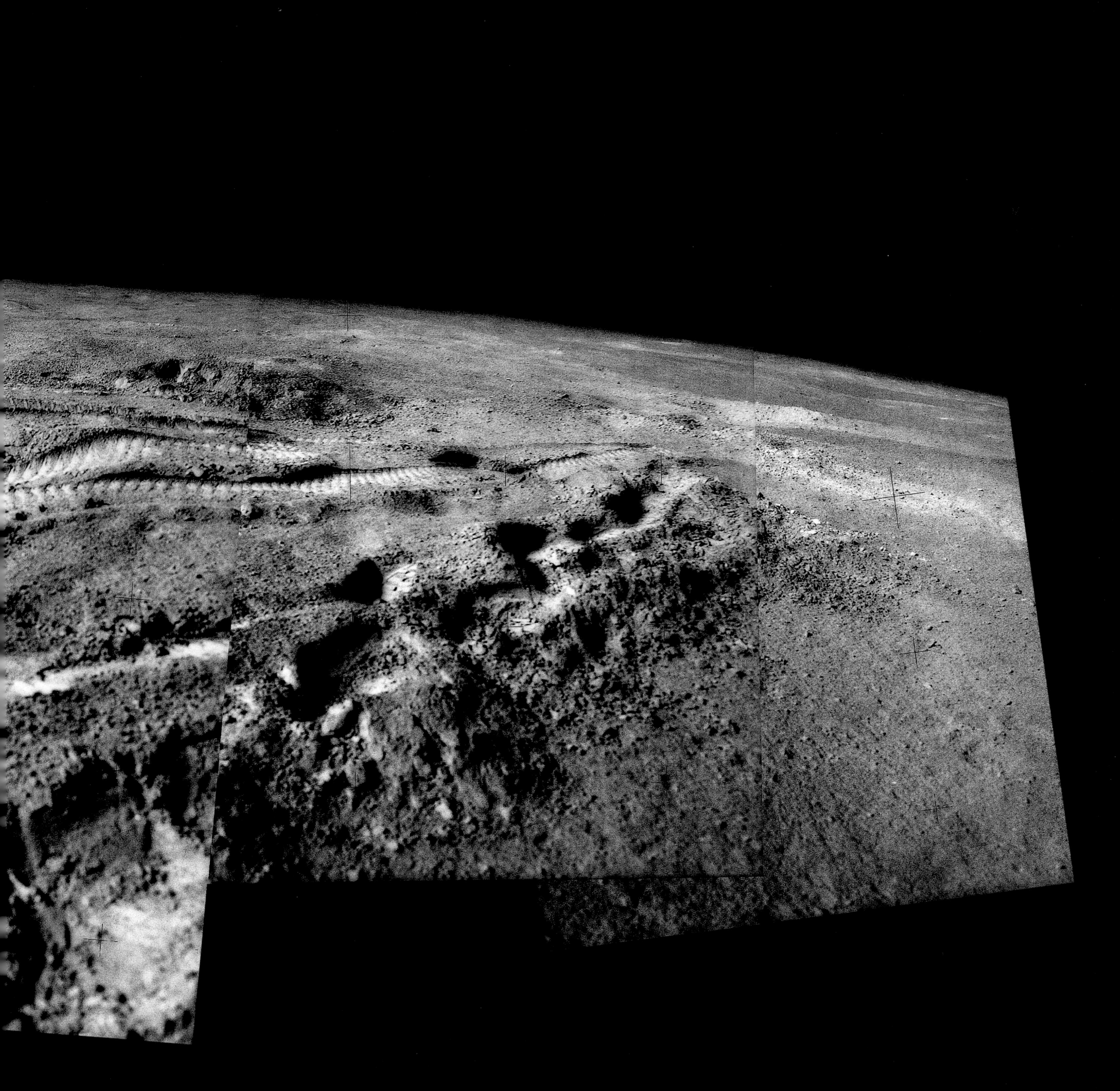

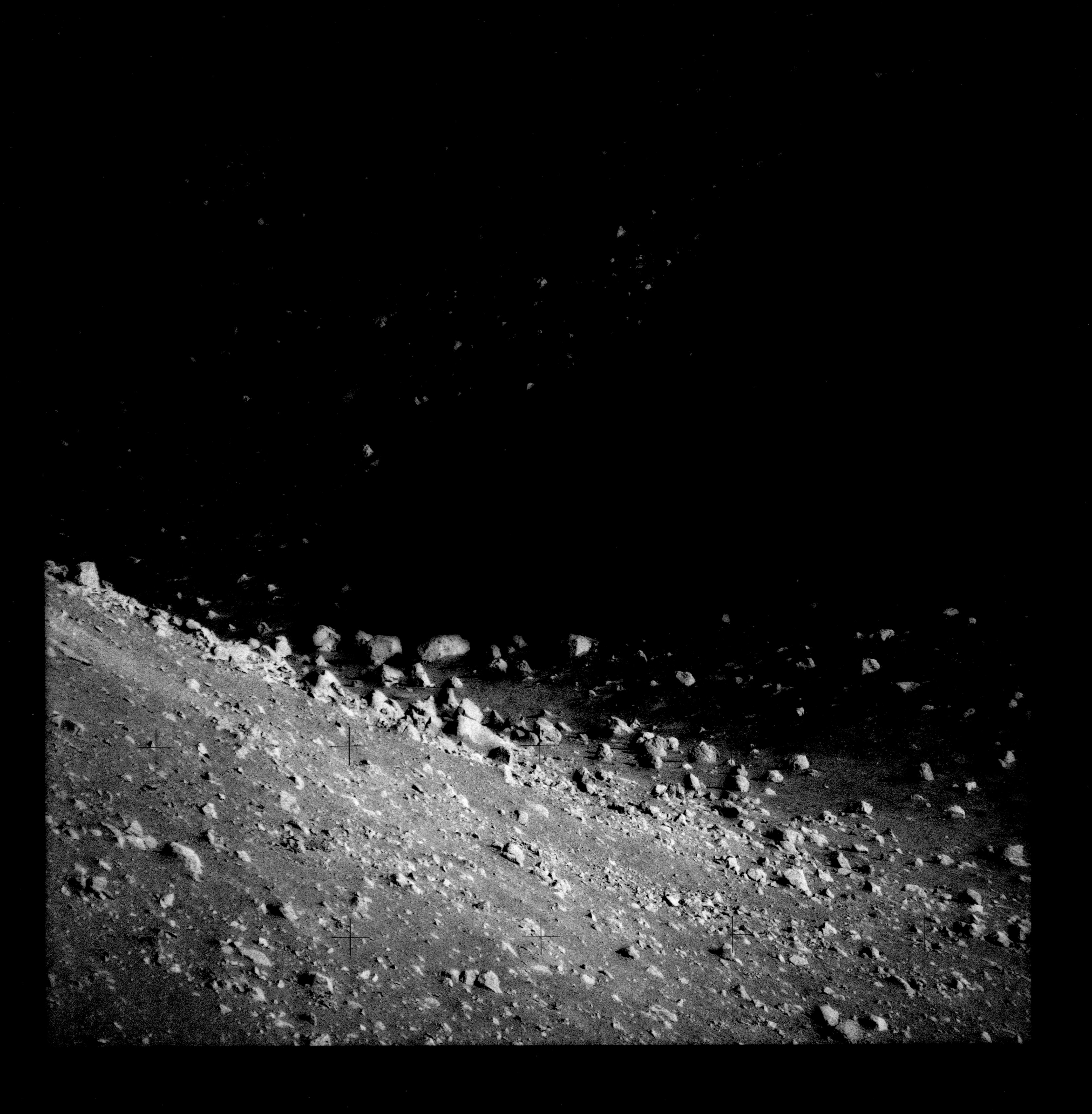

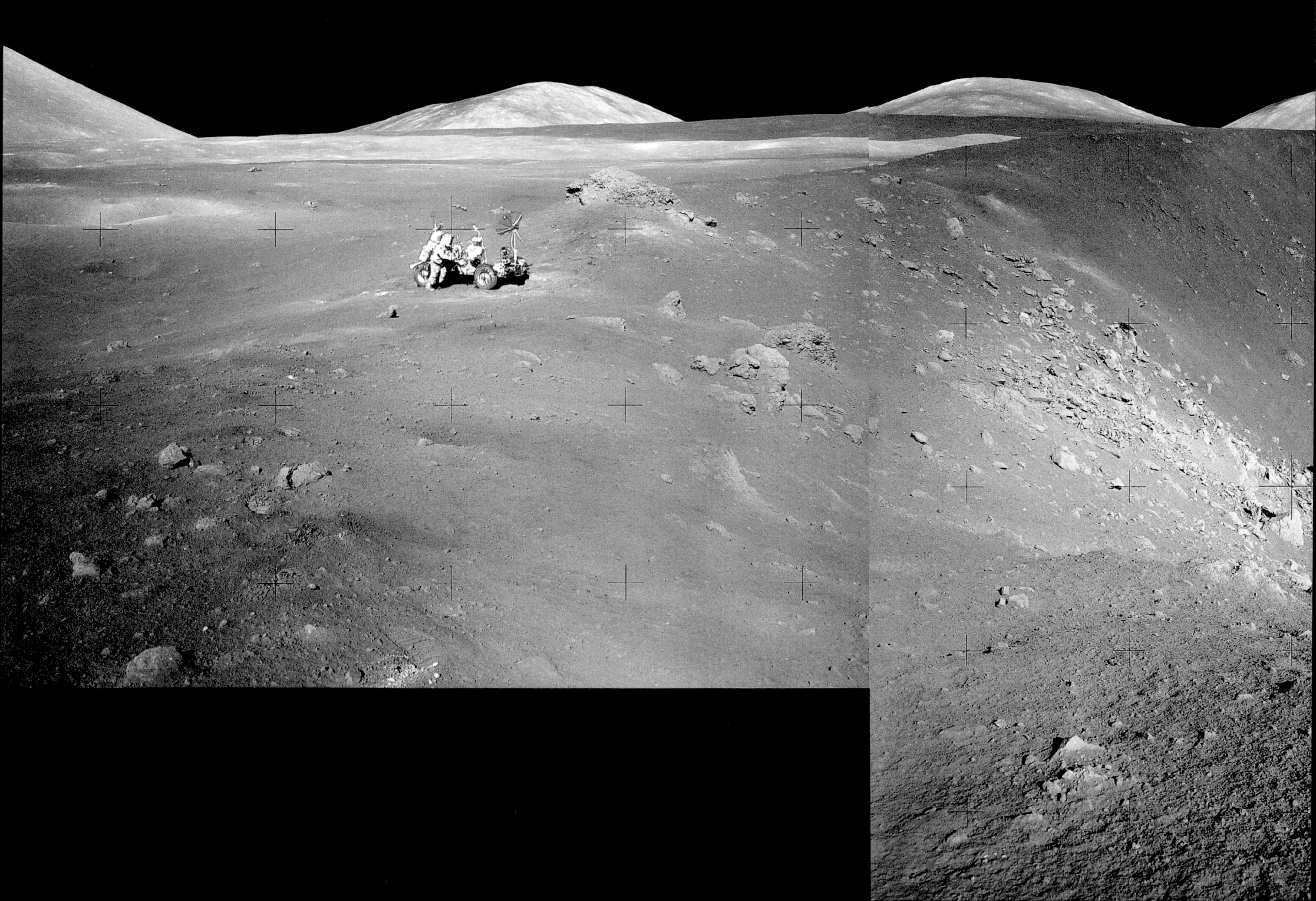

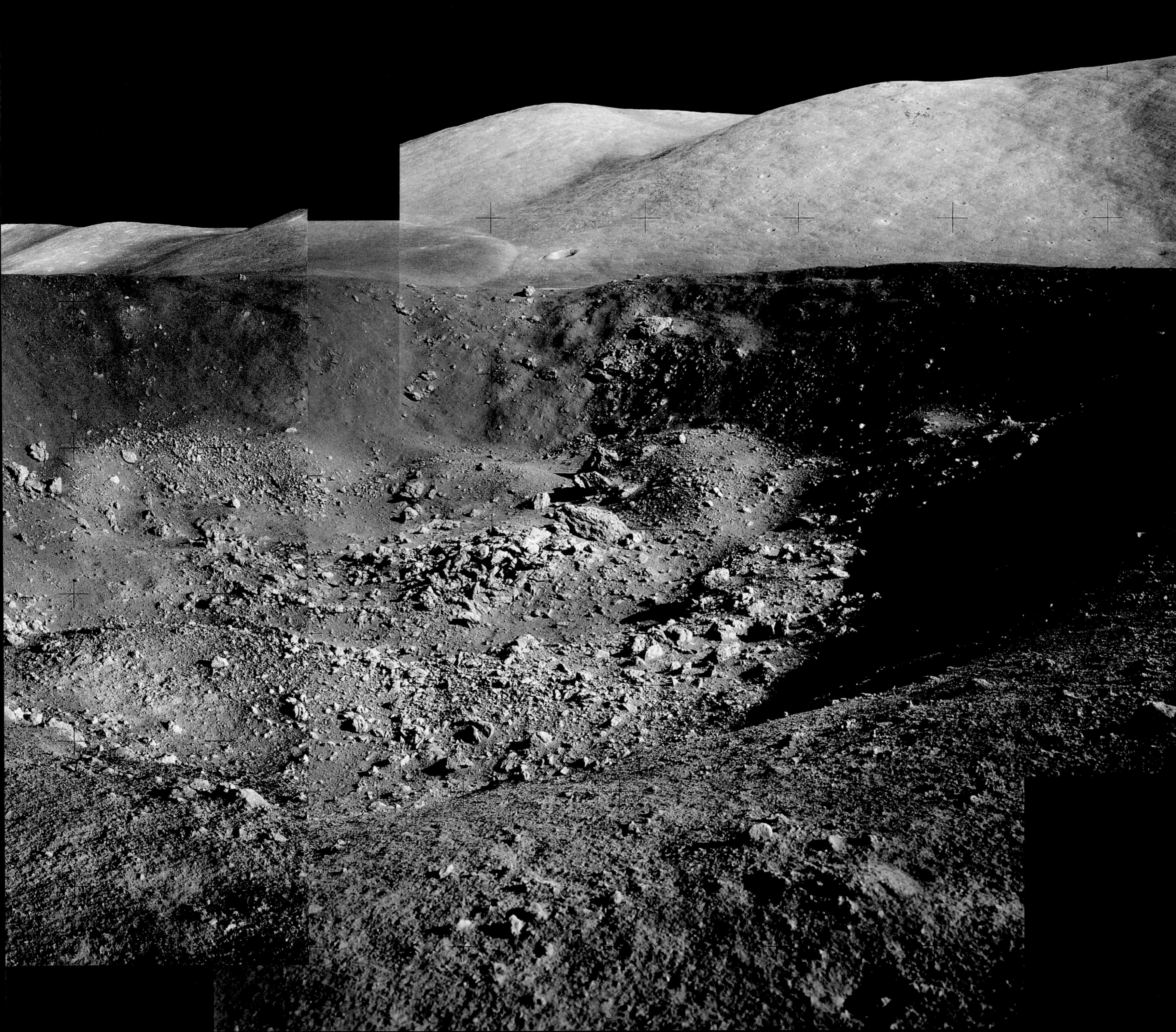

MAN'S FIRST WHEELS ON THE MOON
DELIVERED BY FALCON
JULY 30 1971

UNI
STA

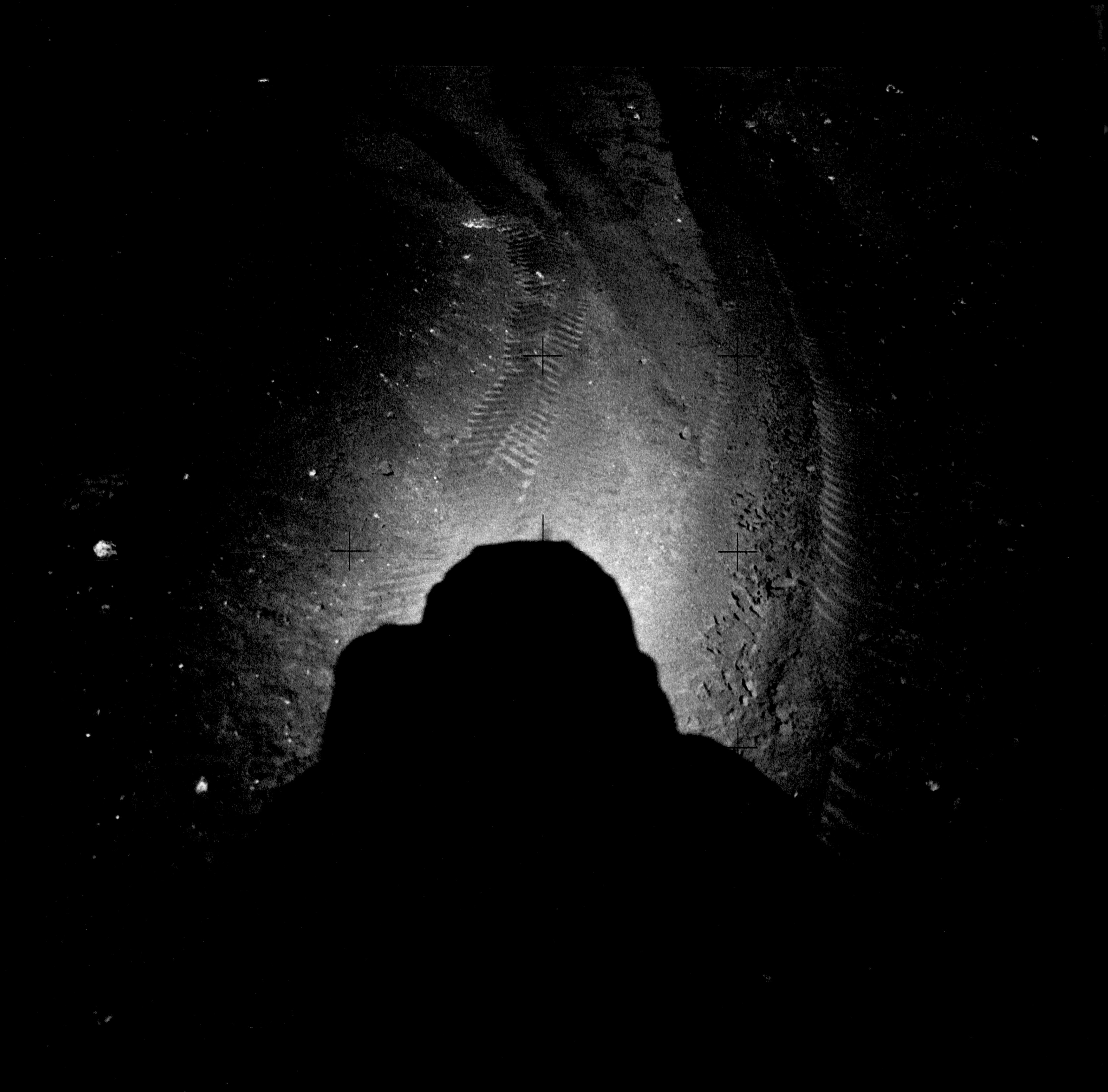

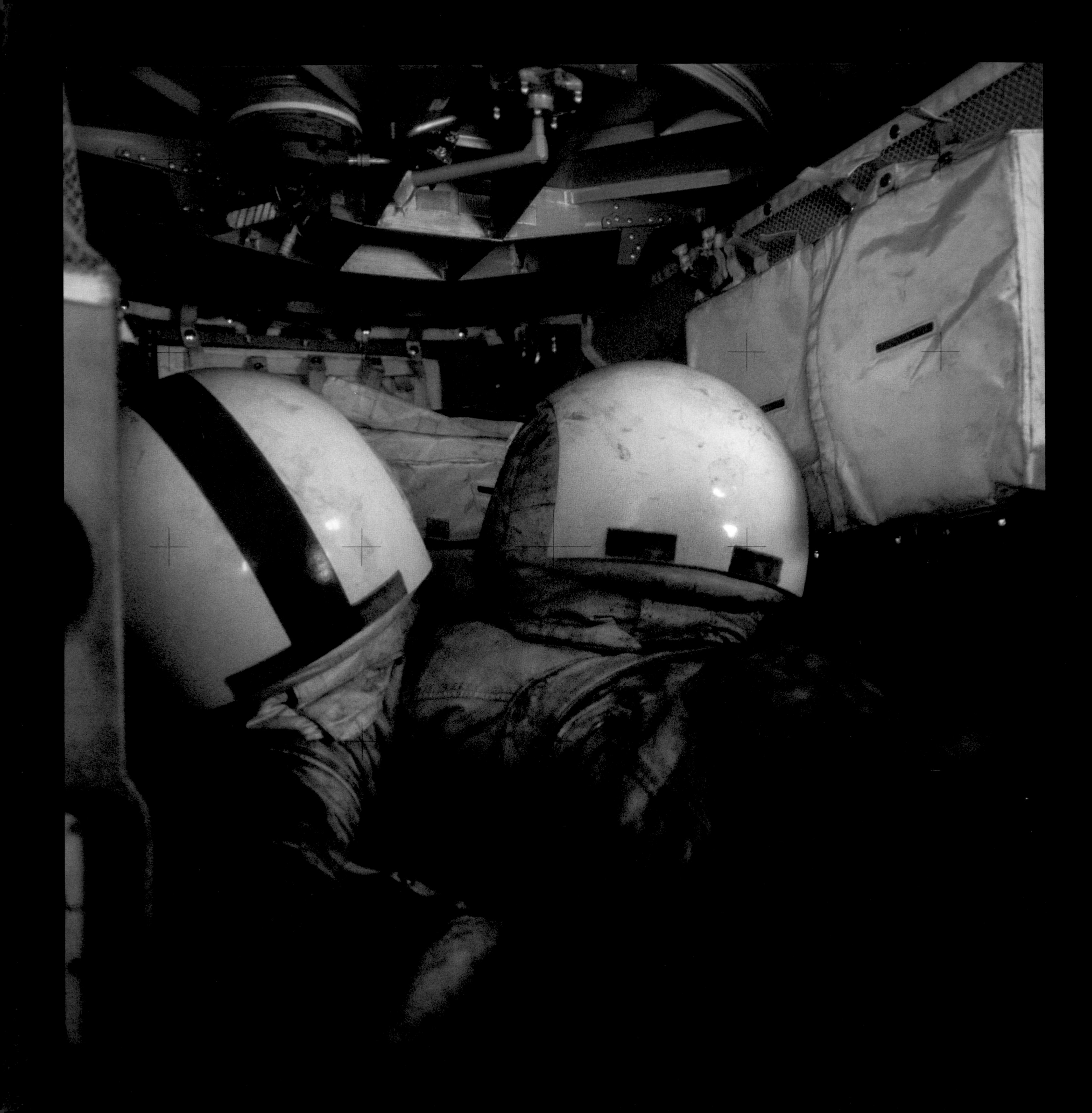

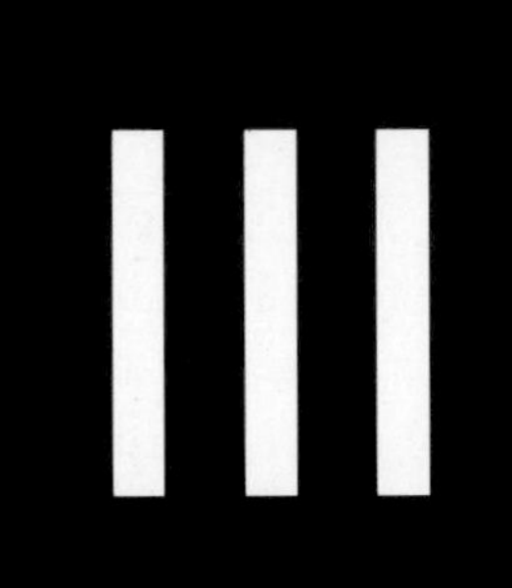

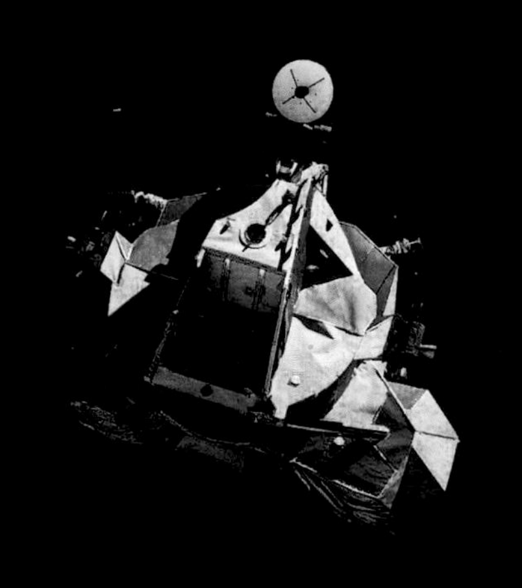

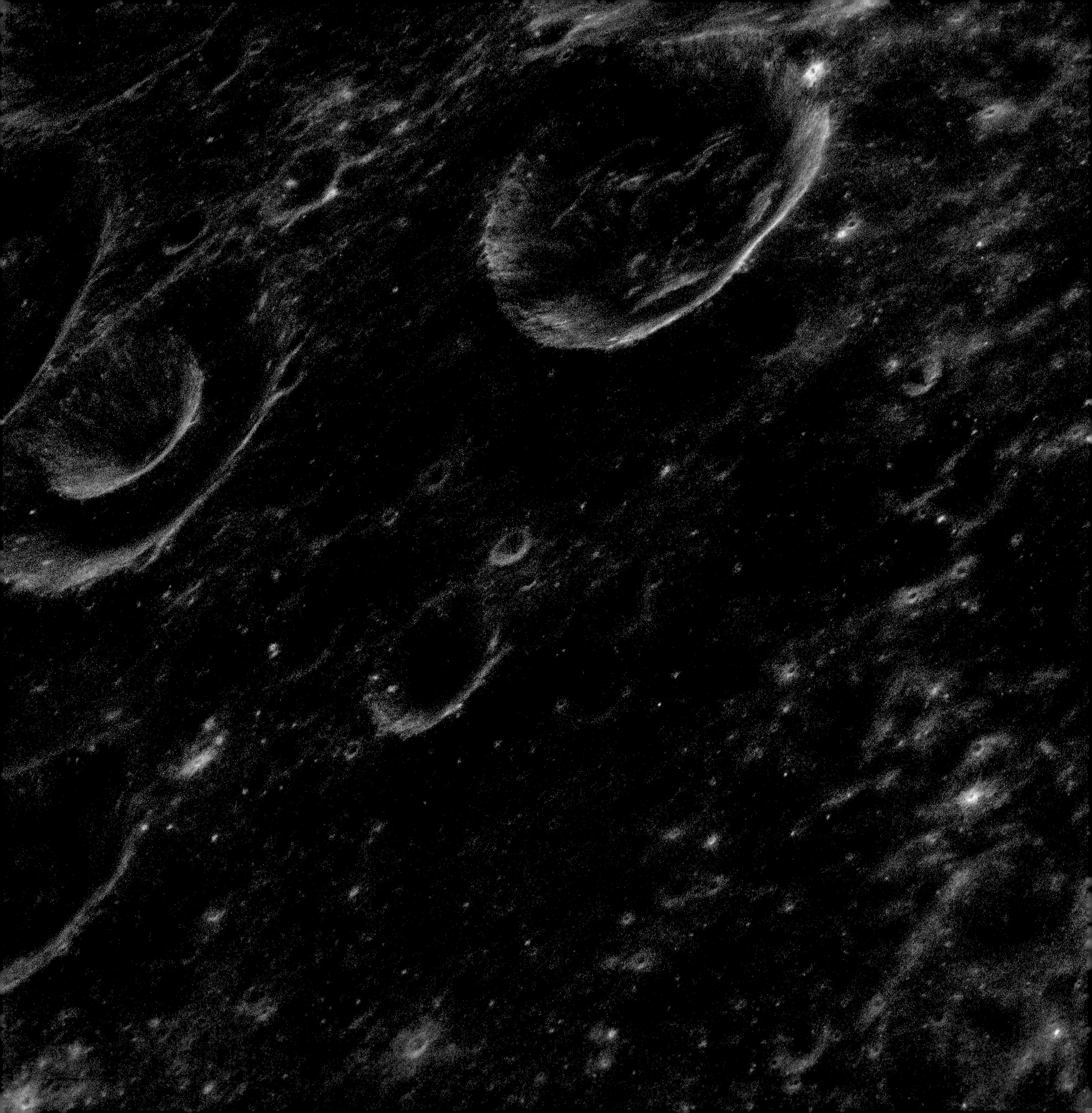

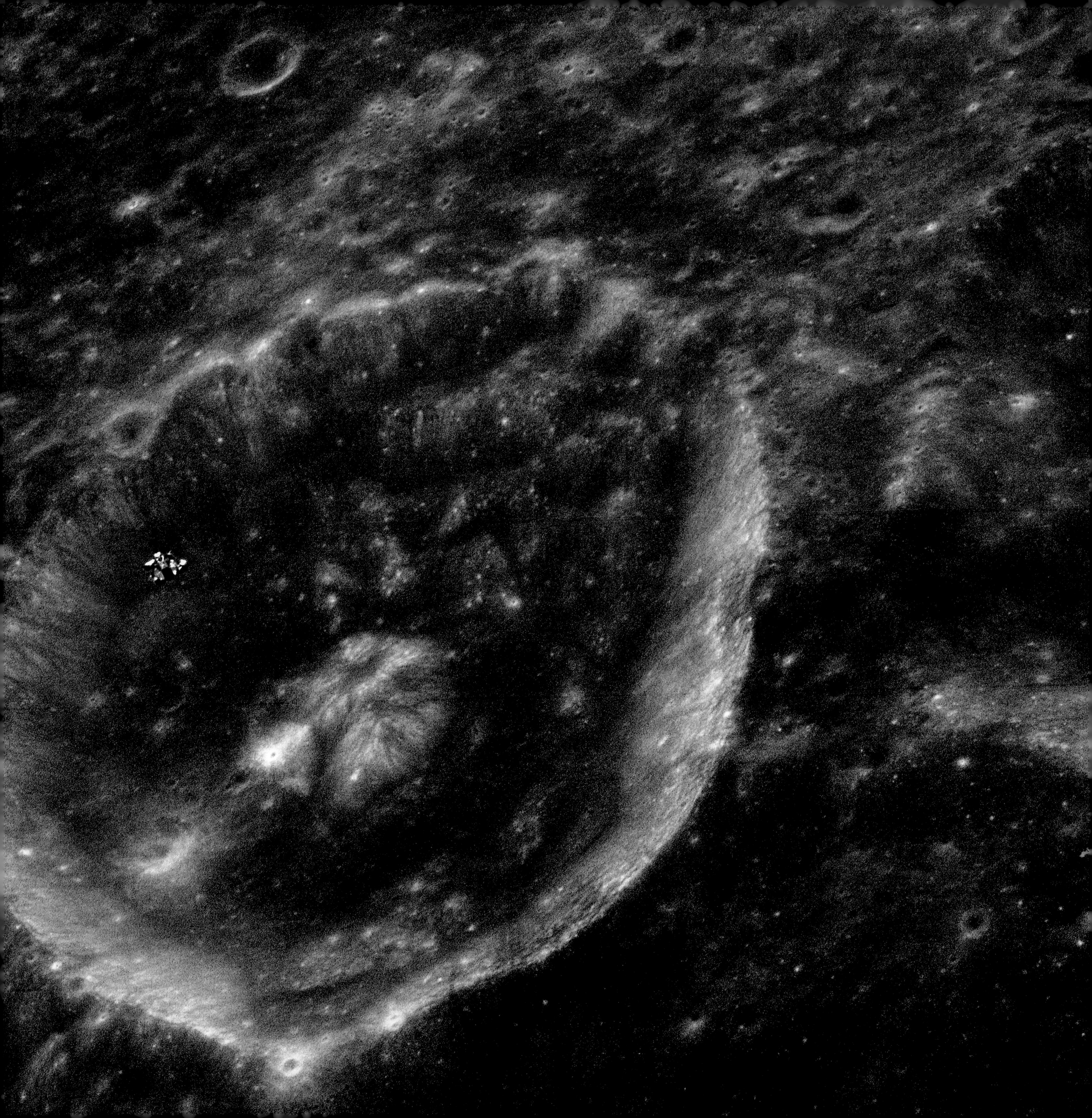

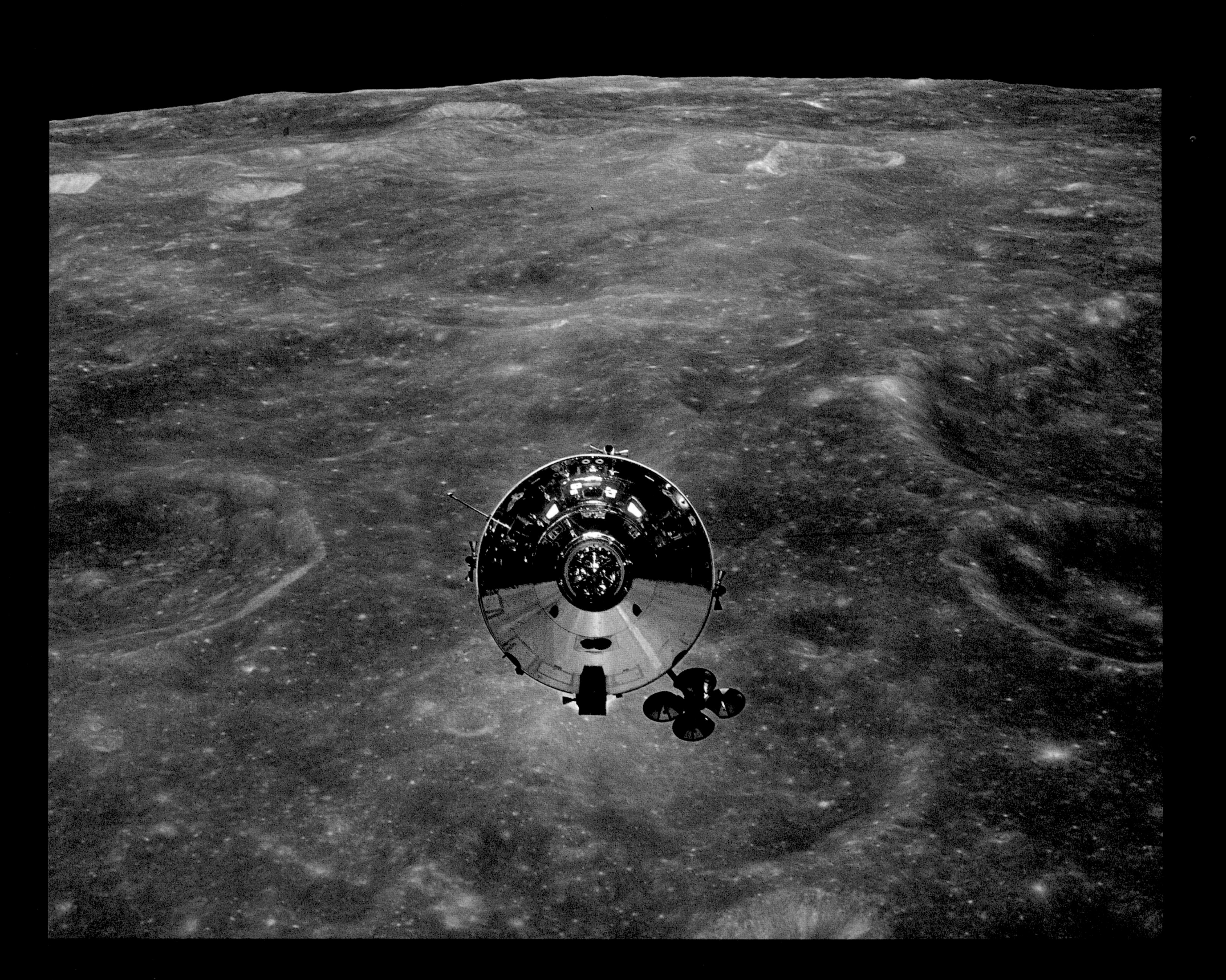

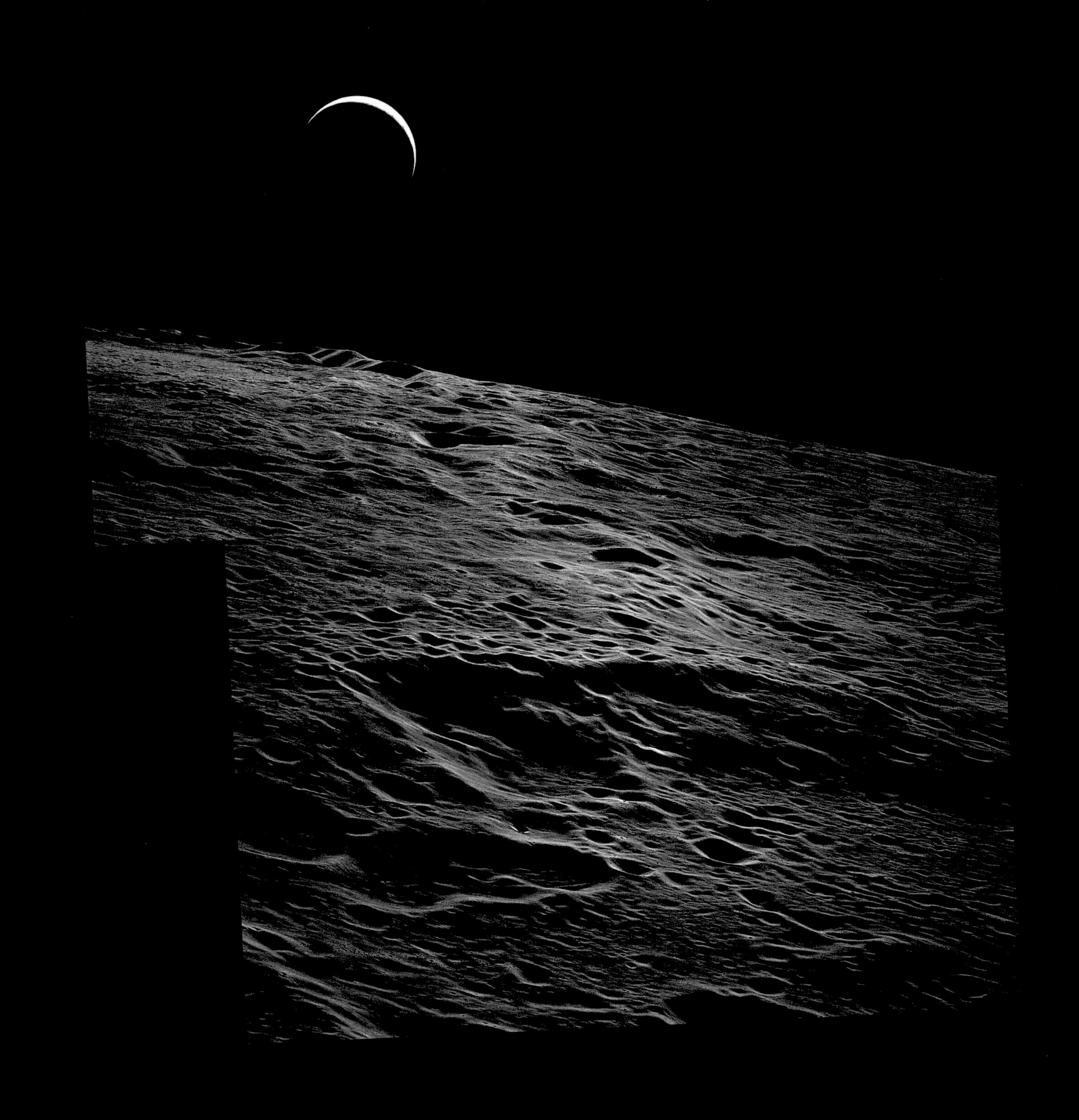

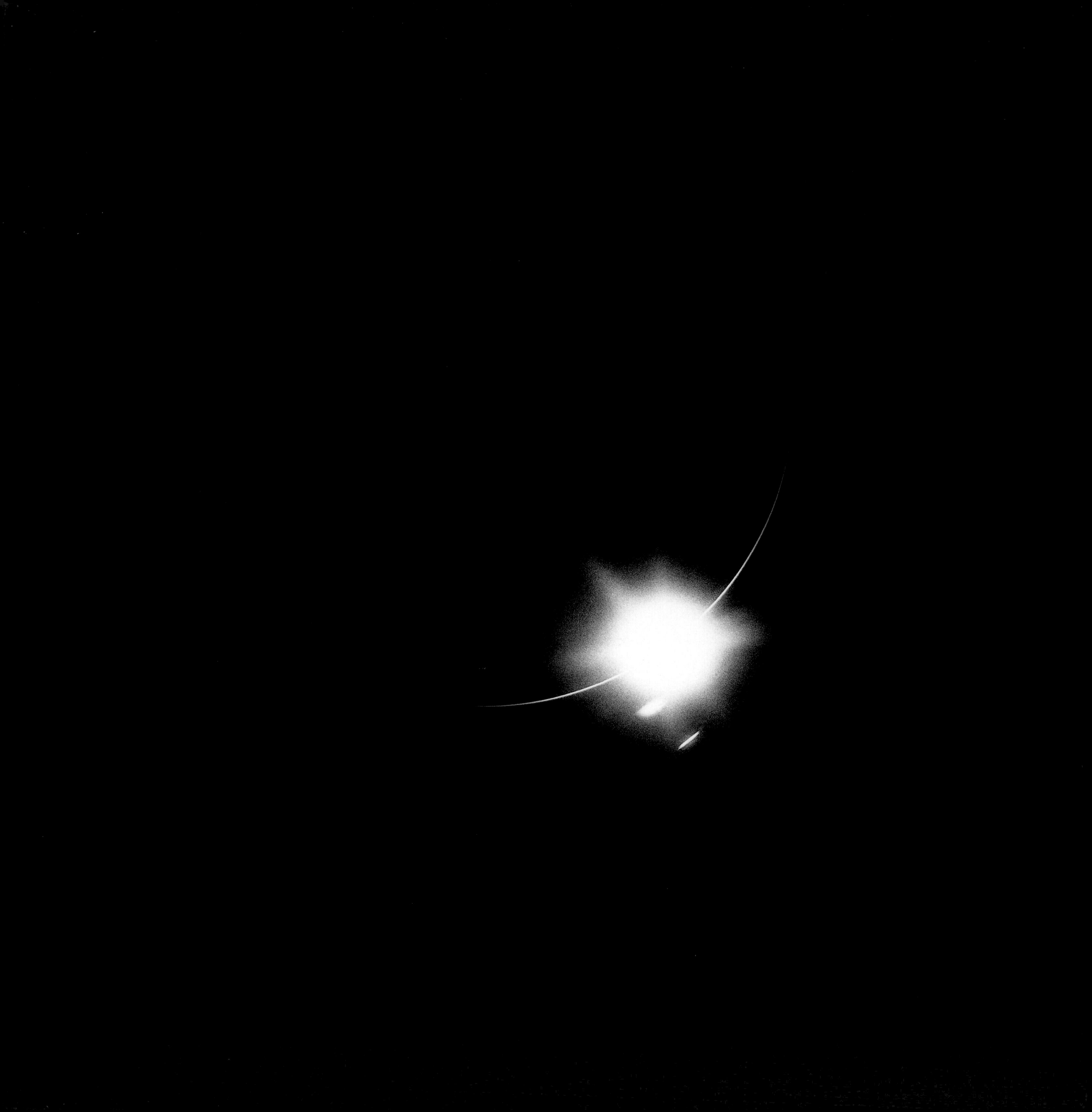

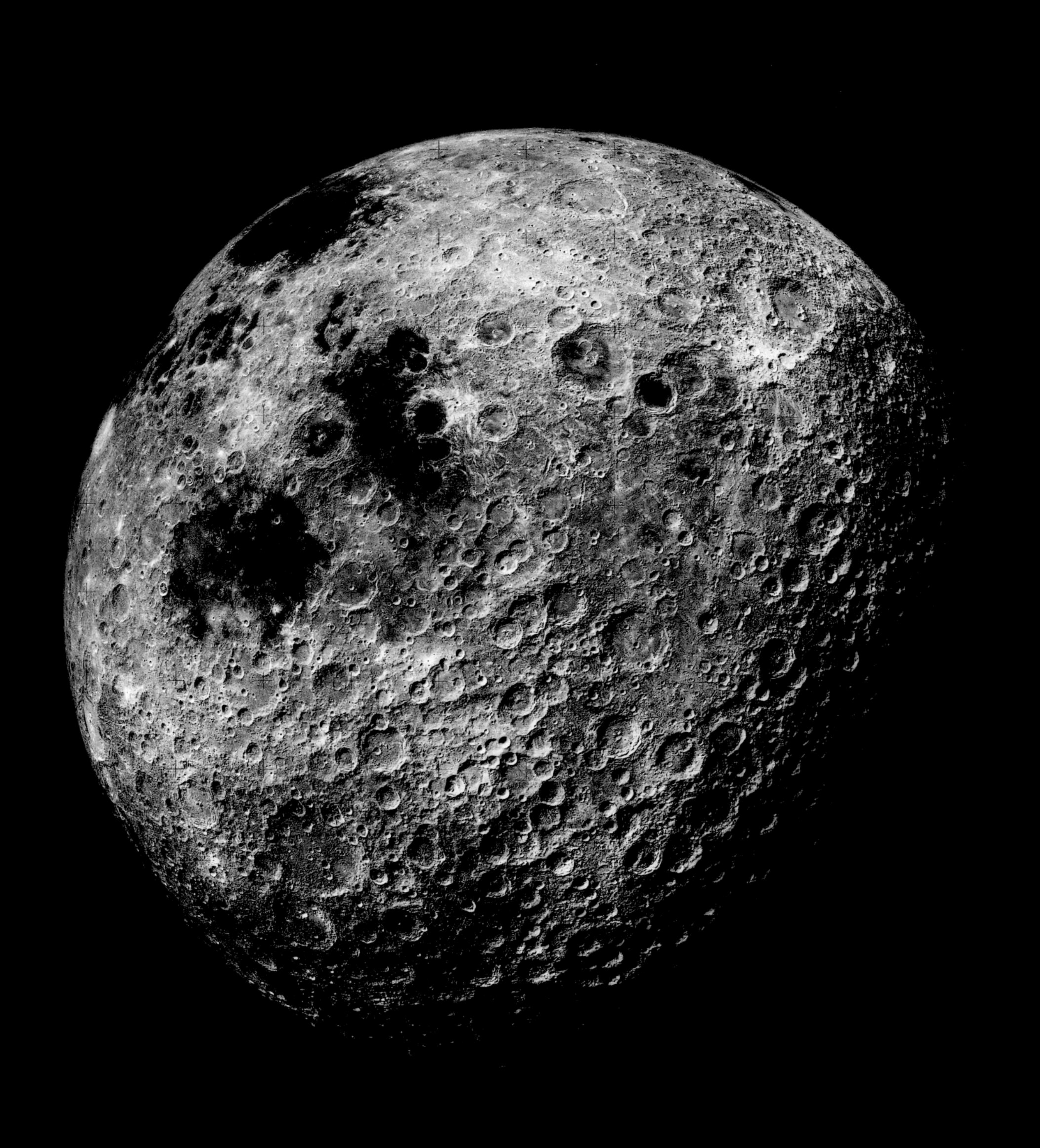

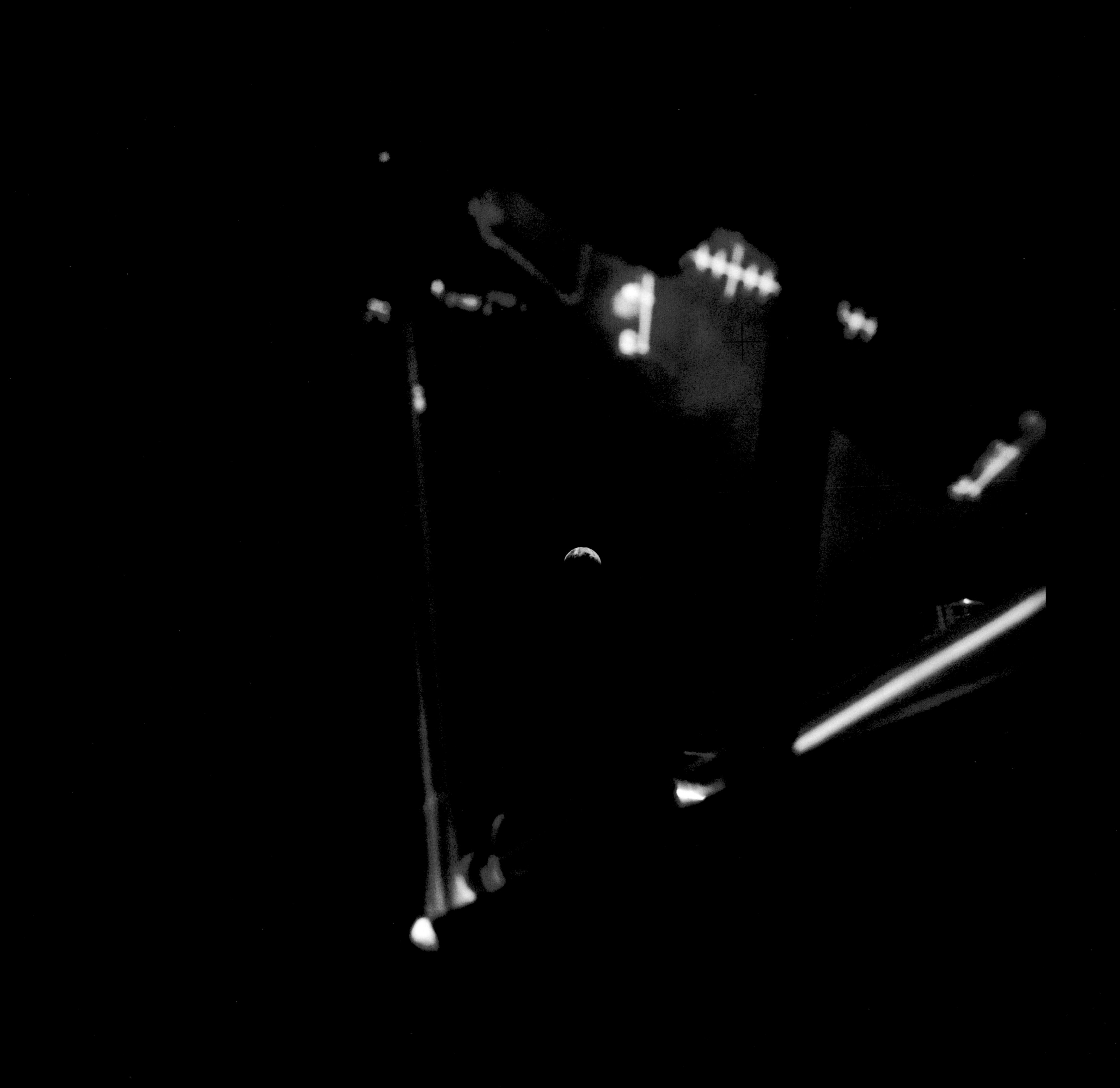

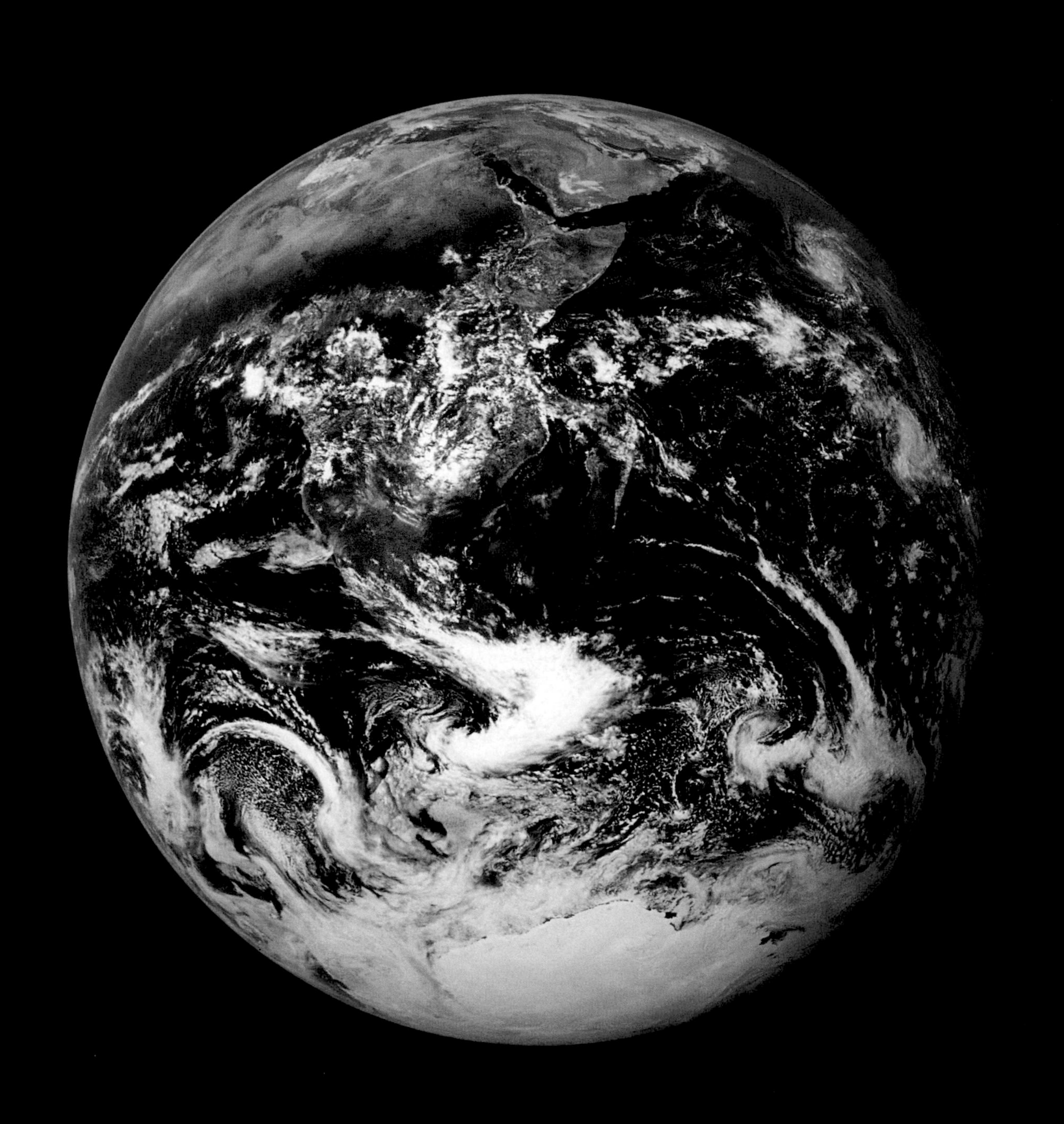

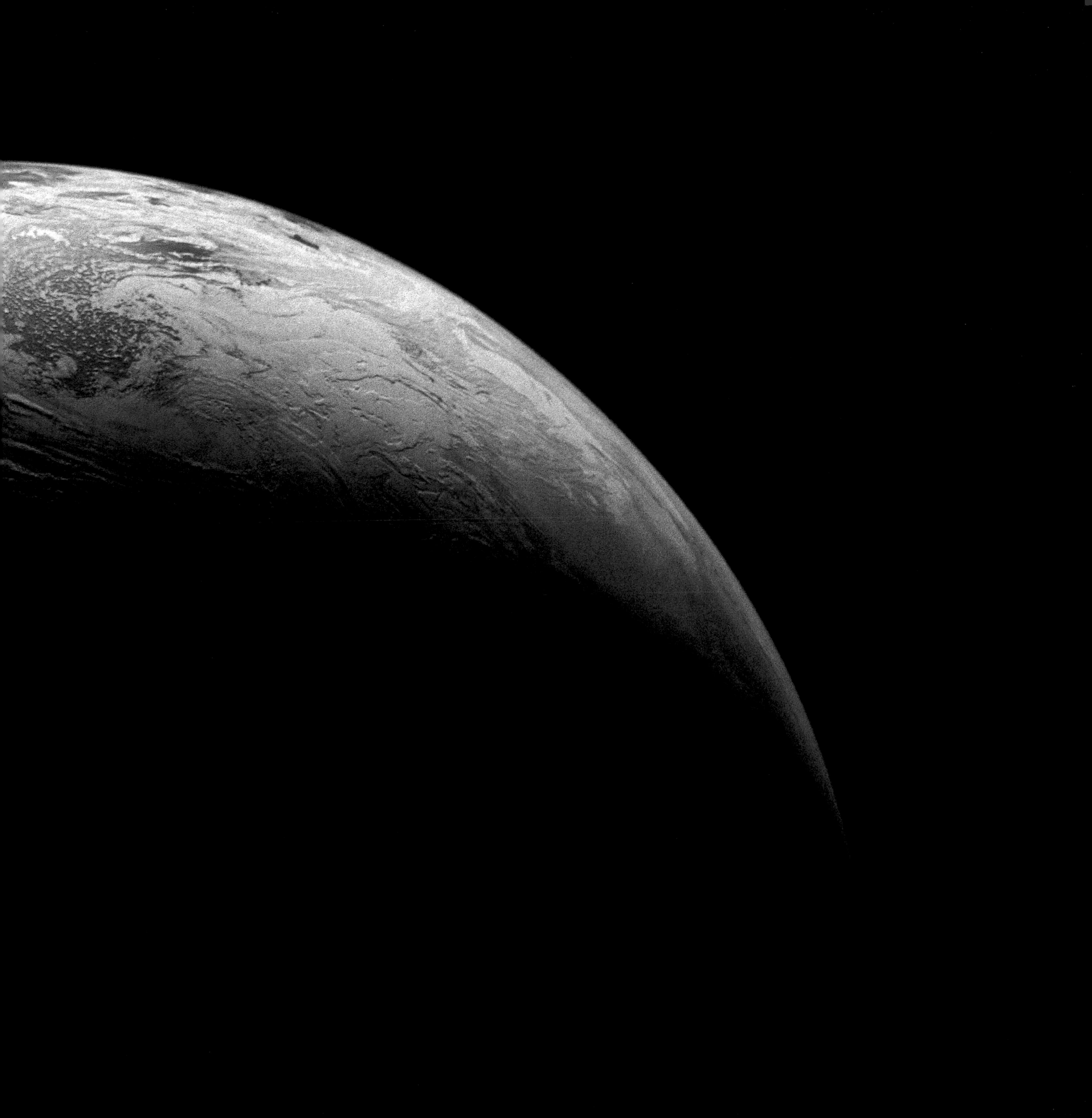

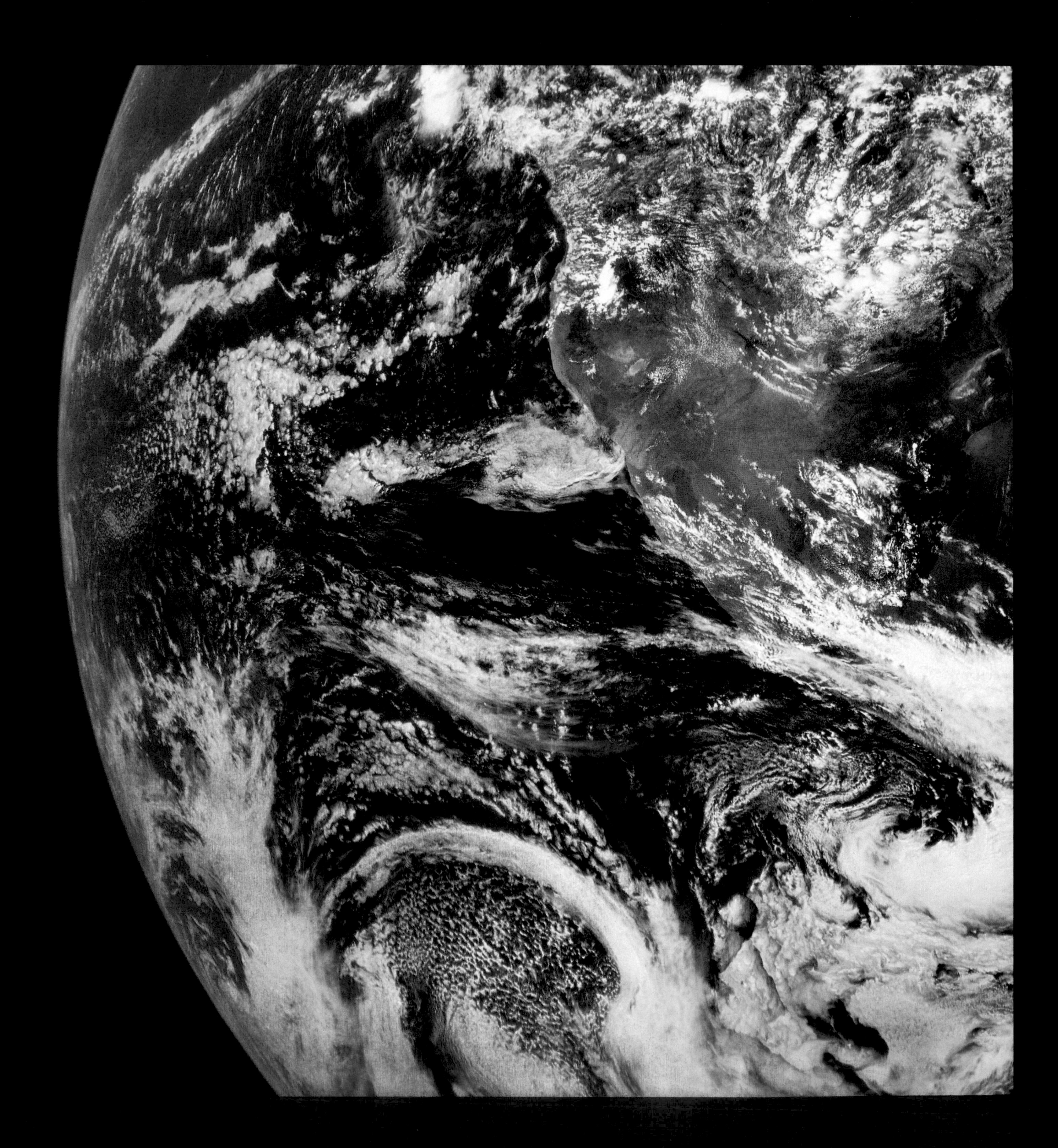

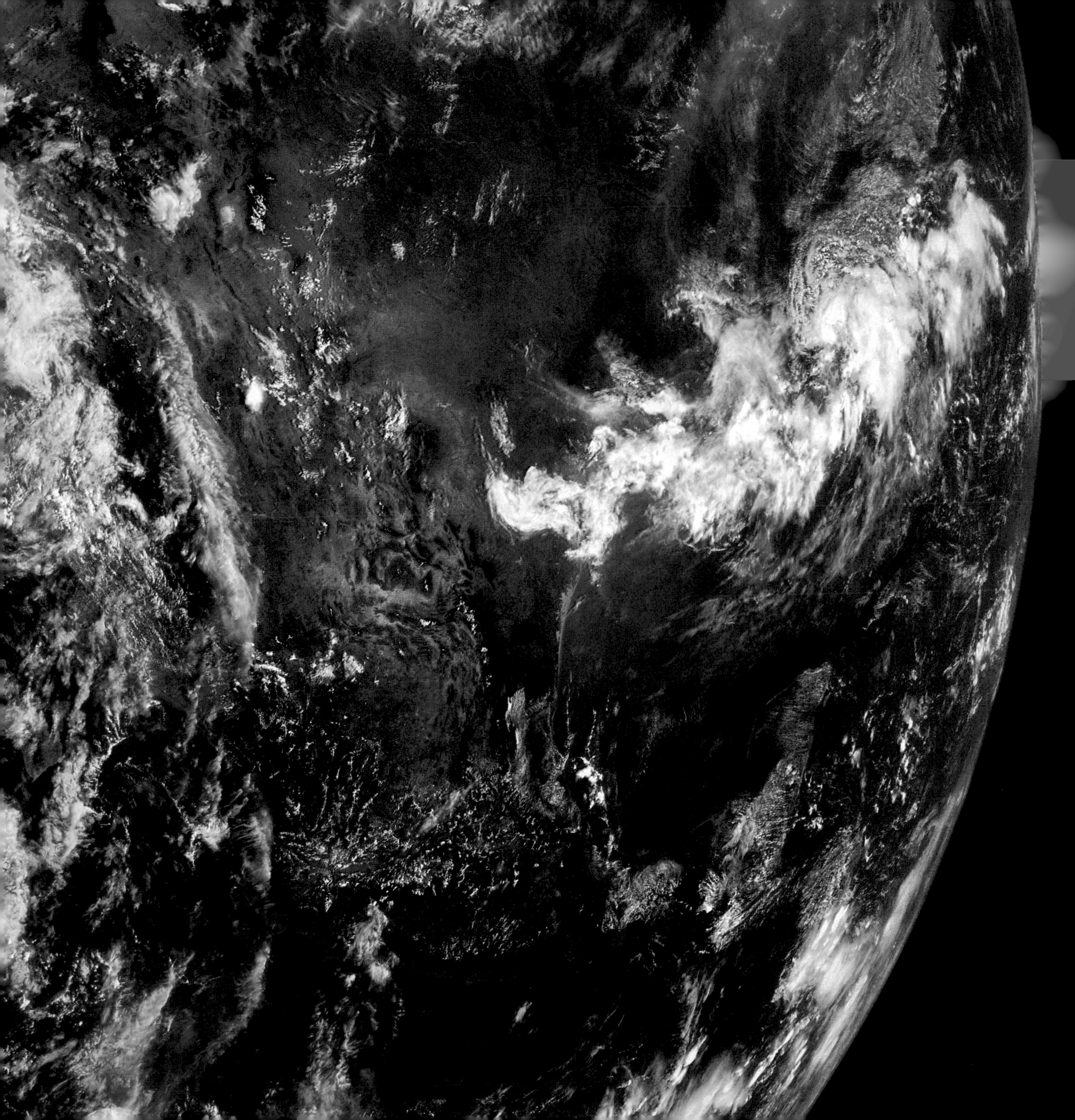

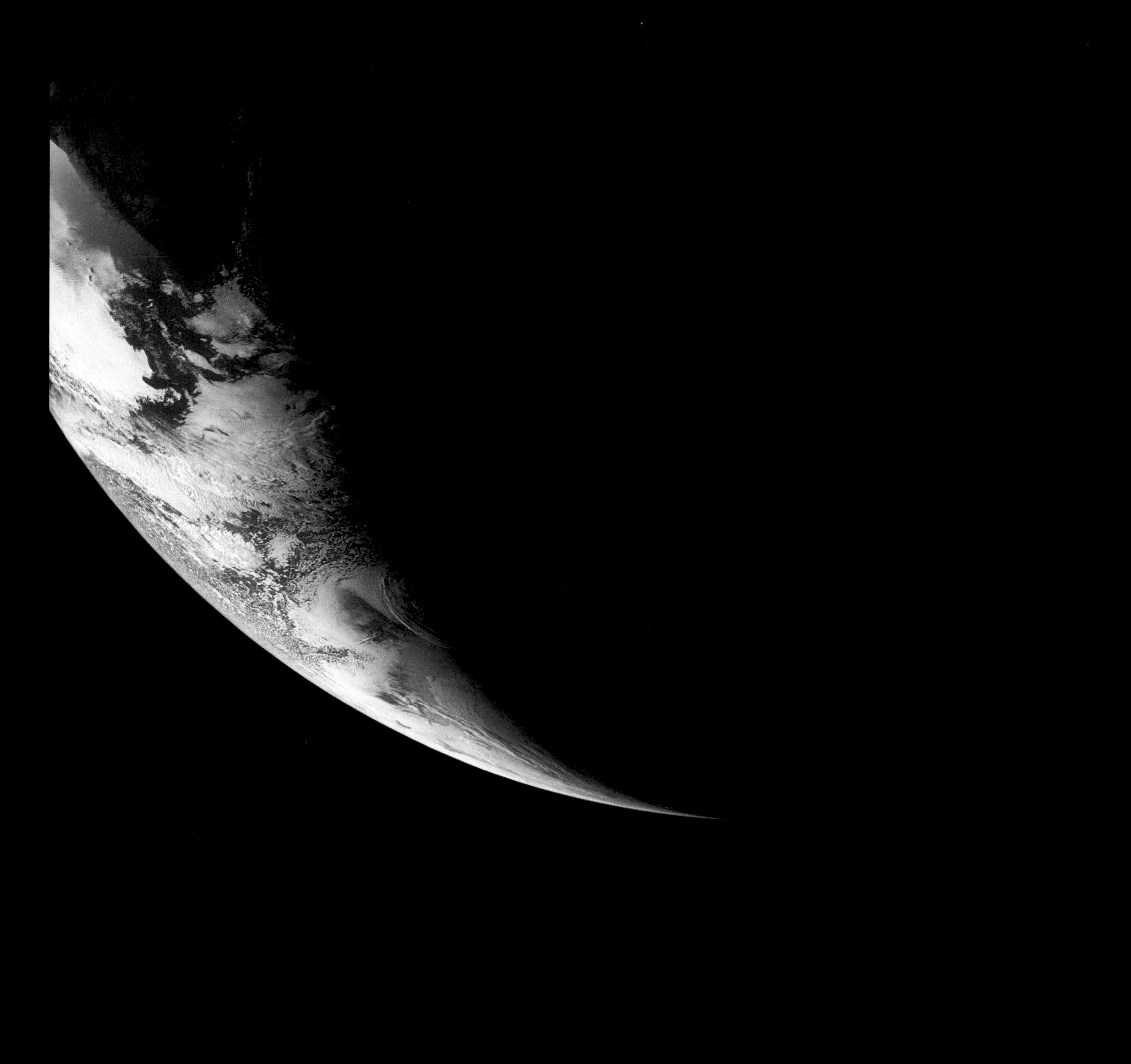

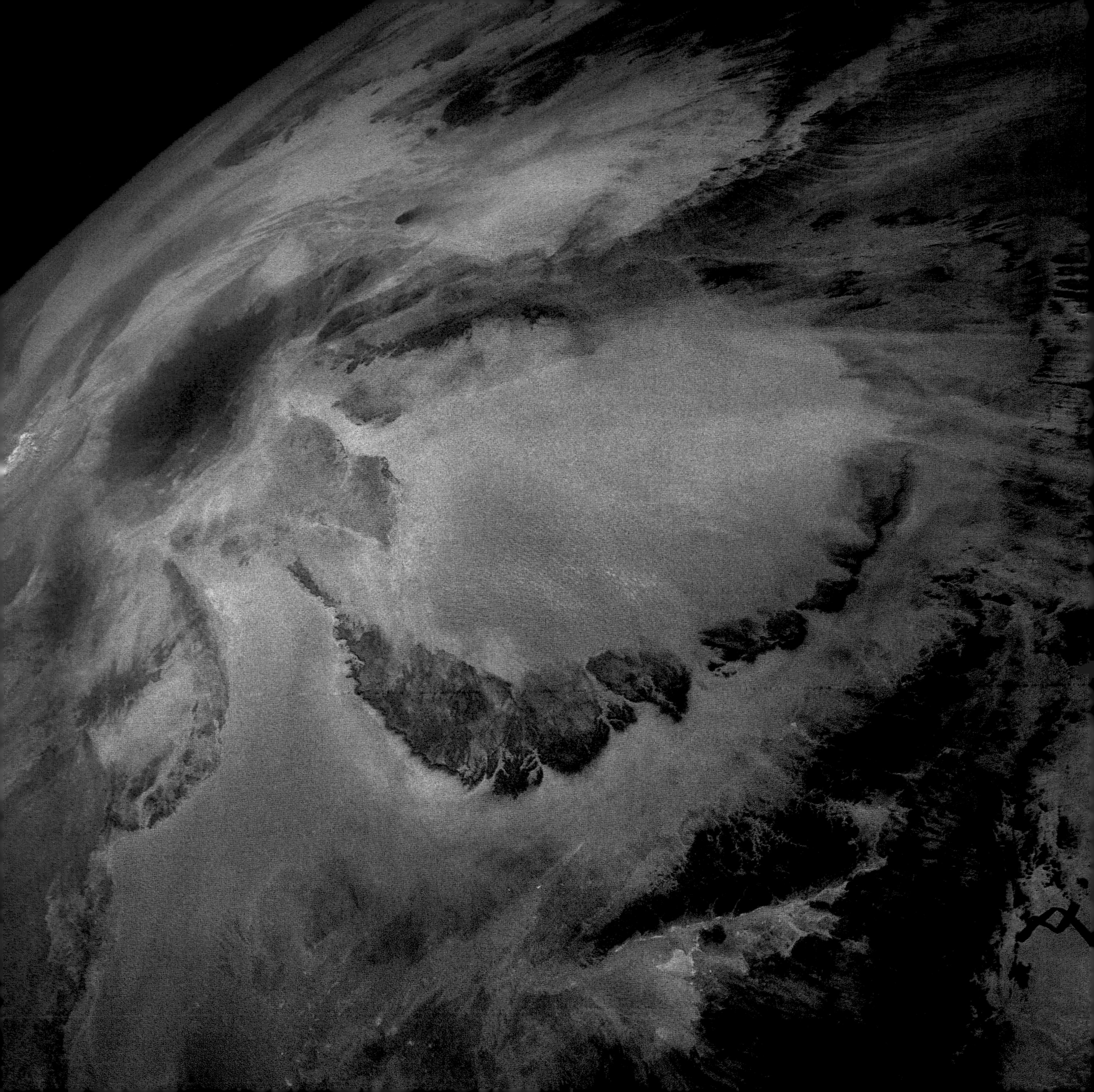

APOLLO MISSION DATA

Apollo 1: January 27, 1967
Crew: Commander (CDR) Virgil I. Grissom, Command Module Pilot (CMP) Edward H. White II, Lunar Module Pilot (LMP) Roger B. Chaffee
Description: Crew perished tragically in cockpit fire during pre-launch ground test.

Apollo 7: October 11–22, 1968
Crew: CDR Walter M. Schirra, Jr., CMP Donn F. Eisele, LMP R. Walter Cunningham
Description: 1st manned test of command and service modules, 163 Earth orbits.
Mission duration: 10 days, 20 hrs, 9 min

Apollo 8: December 21–27, 1968
Crew: CDR Frank Borman, CMP James A. Lovell, Jr., LMP William A. Anders
Description: 1st manned flight around Moon; 10 orbits on Christmas Eve, 1968.
Time in lunar orbit: 20 hrs, 7 min
Mission duration: 6 days, 3 hrs, 1 min

Apollo 9: March 3–13, 1969
Crew: CDR James A. McDivitt, CMP David R. Scott, LMP Russell L. Schweickart
Description: Earth-orbit test of the entire Apollo spacecraft. Included rendezvous between command module and lunar module; 38-minute space walk.
Spacecraft (command module, lunar module): *Gumdrop, Spider*
Mission duration: 10 days, 1 hr, 1 min

Apollo 10: May 18–26, 1969
Crew: CDR Thomas P. Stafford, CMP John W. Young, LMP Eugene A. Cernan
Description: Rehearsal for landing; lunar module descended to 50,000 feet above Moon.
Spacecraft (command module, lunar module): *Charlie Brown, Snoopy*
Time in lunar orbit: 2 days, 13 hrs, 41 min
Mission duration: 8 days, 0 hrs, 3 min

Apollo 11: July 16–24, 1969
Crew: CDR Neil A. Armstrong, CMP Michael Collins, LMP Edwin E. Aldrin, Jr.
Description: 1st lunar landing.
Spacecraft (command module, lunar module): *Columbia, Eagle*
Time in lunar orbit: 2 days, 11 hrs, 34 min
Lunar landing date, location: July 20, Sea of Tranquillity
Time on lunar surface: 21 hrs, 36 min
Moonwalk duration: 2 hrs, 31 min
Pounds of samples collected: 47.7
Mission duration: 8 days, 3 hrs, 18 min

Apollo 12: November 14–24, 1969
Crew: CDR Charles Conrad, Jr., CMP Richard F. Gordon, Jr., LMP Alan L. Bean
Description: 2nd lunar landing; 600 feet from unmanned Surveyor 3 probe.
Spacecraft (command module, lunar module): *Yankee Clipper, Intrepid*
Time in lunar orbit: 3 days, 17 hrs, 2 min
Lunar landing date, location: November 19, Ocean of Storms
Time on lunar surface: 1 day, 7 hrs, 31 min
Moonwalk durations: 1st: 3 hrs, 56 min; 2nd: 3 hrs, 49 min
Pounds of samples collected: 75.7
Mission duration: 10 days, 4 hrs, 36 min

Apollo 13: April 11–17, 1970
Crew: CDR James A. Lovell, Jr., CMP John L. Swigert, Jr., LMP Fred W. Haise, Jr.
Description: 3rd landing attempt; aborted following explosion of oxygen tank inside service module; classified as a "successful failure" because of crew rescue.
Spacecraft (command module, lunar module): *Odyssey, Aquarius*
Mission duration: 5 days, 22 hrs, 54 min

Apollo 14: January 31–February 9, 1971
Crew: CDR Alan B. Shepard, Jr., CMP Stuart A. Roosa, LMP Edgar D. Mitchell
Description: 3rd landing, 1st mission devoted entirely to scientific exploration.
Spacecraft (command module, lunar module): *Kitty Hawk, Antares*
Time in lunar orbit: 2 days, 18 hrs, 40 min
Lunar landing date, location: February 5, Fra Mauro Highlands
Time on lunar surface: 1 day, 9 hrs, 30 min
Moonwalk durations: 1st: 4 hrs, 47 min; 2nd: 4 hrs, 34 min
Pounds of samples collected: 94.4
MIssion duration: 9 days, 0 hrs, 2 min

Apollo 15: July 26–August 7, 1971
Crew: CDR David R. Scott, CMP Alfred M. Worden, LMP James B. Irwin
Description: 4th landing, 1st "J-mission" expedition, featuring extended lunar stay time, long-duration backpacks, and the electric, 4-wheel-drive lunar rover car.
Spacecraft (command module, lunar module): *Endeavor, Falcon*
Time in lunar orbit: 6 days, 1 hr, 17 min
Lunar landing date, location: July 30, Hadley–Apennine Plains
Time on lunar surface: 2 days, 18 hrs, 54 min
Moonwalk durations: 1st: 6 hrs, 32 min; 2nd: 7 hrs, 12 min; 3rd: 4 hrs, 49 min
Pounds of samples collected: 169
Mission duration: 12 days, 7 hrs, 12 min

Apollo 16: April 16–27, 1972
Crew: CDR John W. Young, CMP T. Kenneth Mattingly II, LMP Charles M. Duke, Jr.
Description: 5th landing; exploration of the Moon's central highlands.
Spacecraft (command module, lunar module): *Casper, Orion*
Time in lunar orbit: 5 days, 5 hrs, 53 min
Lunar landing date, location: April 20, Descartes Highlands
Time on lunar surface: 2 days, 23 hrs, 2 min
Moonwalk durations: 1st: 7 hrs, 11 min; 2nd: 7 hrs, 23 min; 3rd: 5 hrs, 40 min
Pounds of samples collected: 208.3
Mission duration: 11 days, 1 hr, 51 min

Apollo 17: December 7–19, 1972
Crew: CDR Eugene A. Cernan, CMP Ronald E. Evans, LMP Harrison H. Schmitt
Description: 6th and final landing. Schmitt was first scientist on the Moon.
Spacecraft (command module, lunar module): *America, Challenger*
Time in lunar orbit: 6 days, 3 hrs, 48 min
Lunar landing date, location: December 11, Taurus-Littrow Valley
Time on lunar surface: 3 days, 2 hrs, 59 min
Moonwalk durations: 1st: 7 hrs, 11 min; 2nd: 7 hrs, 36 min; 3rd: 7 hrs, 15 min
Pounds of samples collected: 243.1
Mission duration: 12 days, 13 hrs, 51 min

(Source: Chaikin, ***A Man On The Moon***)

THERE . . .

1. Saturn V rocket lifts off.
2. First stage separates; second stage ignites.
3. Second stage separates; third stage ignites.
4. Earth "parking" orbit.
5. Third stage reignites for "translunar injection."
6. Command Service Module (CSM) separates from third stage.
7. CSM docks with Lunar Module (LM); separates from third stage.
8. Midcourse trajectory correction, if required.
9. CSM engine slows spacecraft into lunar orbit.
10. Two astronauts move to LM, one remains CSM.
11. CSM and LM separate.
12. LM descent rocket fires.
13. Touchdown on lunar surface.

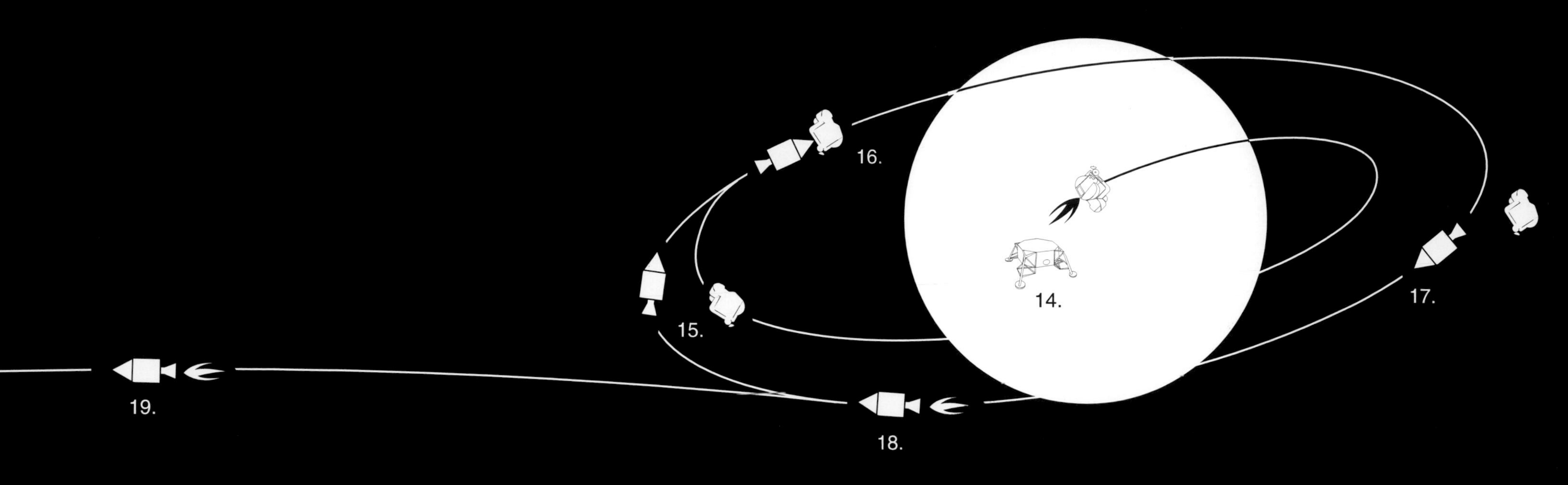

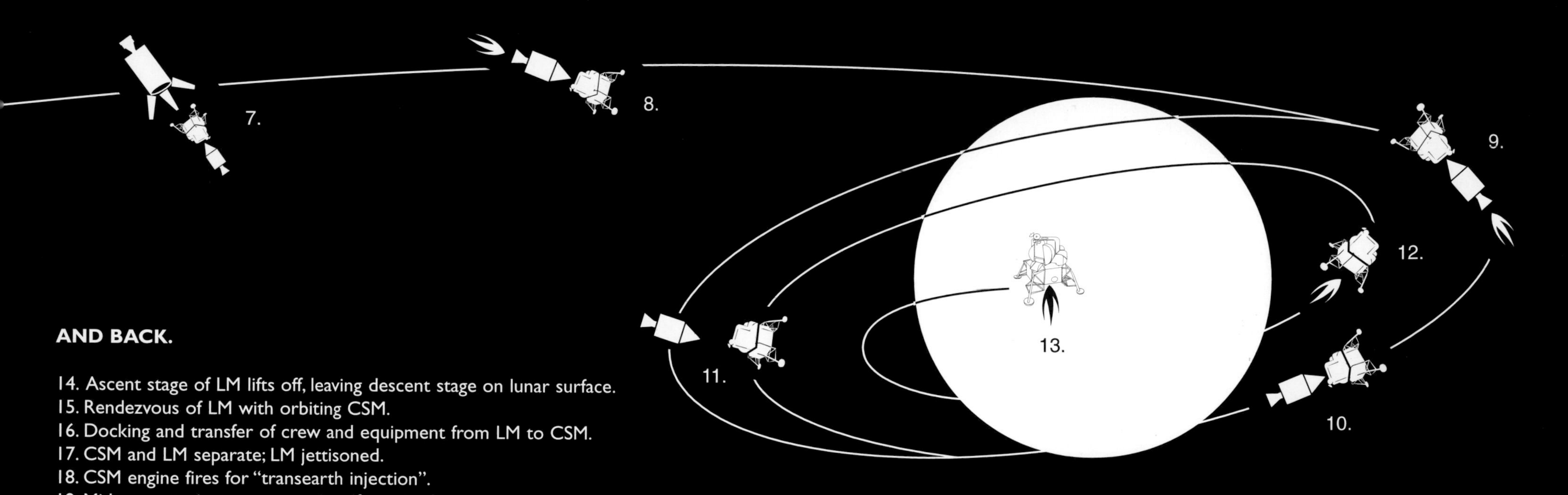

AND BACK.

14. Ascent stage of LM lifts off, leaving descent stage on lunar surface.
15. Rendezvous of LM with orbiting CSM.
16. Docking and transfer of crew and equipment from LM to CSM.
17. CSM and LM separate; LM jettisoned.
18. CSM engine fires for "transearth injection".
19. Midcourse trajectory correction, if required.
20. CM separates from SM; SM jettisoned.
21. Heat shield forward, CM reenters Earth atmosphere.
22. Parachutes deploy.
23. Splashdown in Pacific Ocean.

15
17
11
12
14
16

THE FARTHEST PLACE

They had done dangerous things before. Some had flown in combat. Others had landed a jet fighter on the deck of an aircraft carrier in the middle of the night in the open ocean. Almost all had pushed an unproven supersonic aircraft to the edge of its capabilities. And some, the veterans, had even ridden a rocket into space before. So it was not a completely novel sensation that greeted the three astronauts on launch morning, as they donned their space suits, left the crew quarters, and climbed into the transfer van for the ride to the pad.

And yet, as they stood at the base of the mammoth white rocket called the Saturn V, they knew this flight was literally above and beyond anything they had ever done. Longer than a football field – 363 feet from base to tip – the Saturn was the most powerful rocket ever successfully flown. Loaded with supercold rocket fuel and liquid oxygen, it weighed 6 million pounds and harbored as much explosive energy as an atomic bomb. Among its estimated 3 million parts was an electronic brain programmed to steer the Saturn into Earth orbit. The engine in its third stage would actually be shut down and then reignited, to speed the astronauts on a course for the Moon. The Saturn V was NASA's answer to the Pyramids.

Up on the 320-foot level, a walk across the access arm brought the astronauts to the small White Room. This enclosure nestled against the hatchway of the cone-shaped command module, ferry craft for the lunar voyage. Assisted by technicians, each man slid into the confines of the command module and settled into his couch. Oxygen hoses were hooked up, communications lines connected. A member of the support crew reached inside to clasp the gloved hands of his friends, in a wordless farewell. Then the hatch was closed, sealing the Moon voyagers inside their spacecraft.

Now ended years of anticipation, beginning with their selection as astronauts and building with their growing involvement in the Apollo program, and finally, their selection for a lunar mission. They were not here because they had harbored childhood dreams of going to the Moon; when they were growing up, astronauts were purely science fiction. They were here because they had been born at the right time, and because they were among the best pilots of their generation. They knew, as they lay at the pinnacle of the great rocket, that they were simply the most visible members of the Apollo team: a work force estimated at 400,000 men and women. But they alone would make the journey. Although they were men of mission, and felt fortunate to carry out their country's mission, the lure was mostly personal, professional: to fly as no one had flown before.

As they lay in their couches, setting switches and making checks, the men could almost believe this was just another simulation, like countless others they had logged during the months of training. But as the count reached T-minus 15 minutes, they knew this was no practice run, that today they were really going. Adrenalin began to flow as the pace of events quickened. In the final seconds, the five engines of the Saturn's first stage ignited, their roar audible through command module walls and space suit helmets. At last the Saturn rose from the Earth, its light akin to a second sun, its roar powerful enough to be detected hundreds of miles away. Something as audacious as a trip to the Moon could not begin quietly.

\+ + +

Crossing the 240,000-mile gulf between Earth and Moon took about 66 hours, roughly the time needed for an ocean liner to sail the Atlantic. The astronauts coasted moonward in the lap of celestial mechanics: as Apollo 8's Bill Anders put it during the first lunar voyage, "Isaac Newton is doing most of the driving right now." During that time the astronauts were not lacking for things to do. Spacecraft systems were monitored and maintained. Navigation checks were made, using an onboard sextant to sight on the Sun, Earth, and stars. Every procedure was so familiar, so well practiced, that as long as everything was going well the men could have been in the simulator – except for two things.

One was weightlessness, a continuous sensation of free fall that lasted throughout the voyage, interrupted only by the firing of a rocket engine, and, for the two moonwalkers, the visit to the lunar surface. Weightlessness made the cramped command module seem suddenly roomy; ceiling and floor were equally good places in which to work. An astronaut could move around by his fingertips. A camera hung in mid-air until he reached for it. A glob of water became a shimmering sphere, suspended in the cabin. The men slept, literally, on air.

Then there was the view. In the hours after the so-called Translunar Injection rocket firing by the Saturn's third stage, the astronauts saw their home world dwindle until it looked no bigger than a baseball held at arm's length. It grew smaller by the hour, until it was all but lost in a sea of blackness. The astronauts' radio transmissions, traveling at the speed of light, took a noticeably long moment to reach Mission

THE SKIN OF THE MOON

Like most people alive between 1967 and 1972, I remember the Apollo missions. They began pretty dimly for me, at age four, but by the time Eugene Cernan and Harrison Schmitt concluded the lunar explorations with Apollo 17, this worldly nine-year-old had come to accept men on the Moon as a predictable – if still thrilling – fact. What I remember most from that time are the basics: the constant presence of the television, the brief riveting drama of liftoff, and the interminable waiting and commentary by the pundits and news anchors. The coverage was leavened only infrequently by stilted and grainy inflight video transmissions, and proved overall to be substantially less interesting than the rich fantasy life of the young space adventurer. Neil Armstrong's strange and murky television image setting first foot on the Moon is seared into the world's collective brain, mine included, but more vivid for me are the eerie rhythms of radio communication between the astronauts and ground control in Houston – that inimitable *beeeep*, followed by the most spacious silence imaginable. Those key sounds come back clearly, along with certain visual icons. How can I forget the ghostly Saturn V rocket rising up over Florida on a column of fearful flame, over and over and over again in my parents' living room, as the years passed and they went about their domestic routine? The era ended quickly, just as the world and the boy began to take it for granted, and we all moved on.

Twenty-five years elapsed until I thought much about Apollo again. This time, it was as a photographer with a particular interest in exploration and landscape issues. At a certain point in my own work I had realized that a landscape is a landscape, whether made on Earth, the Moon, or Mars, and had become fascinated with the differences and similarities between images of our home ground and those of other worlds. The Moon held a particular attraction, however. While robotic probes had traveled most of the solar system since 1972 and had radioed their extraordinary images back to Earth throughout, the only celestial realm humans had actually traveled to and photographed in person was the Moon. I was also especially taken with the way the Moon's spare physical characteristics offered a distillation of traditional landscape vocabulary, a kind of pure and elemental dialogue between minerals and radiation, an epiphany of rock and light.

Certainly my interest had shifted from the boyish fascination of years past, but deep down I must admit to being driven by essentially the same forces: I still wanted to be an astronaut. More precisely, I suppose, I wanted to be an extraterrestrial photographer. Needless to say, I faced some formidable obstacles to realizing such an ambition, so I settled for investigating the next-best thing available, the inflight photographs that the Apollo astronauts had already made. The pictures I was familiar with were the twenty or so well-worn images that we all know, the iconic photographs that were selected at the time by the editors of magazines like *Life* and *National Geographic*, and that have come to define a certain part of late-twentieth-century visual culture. No pictures are more famous – Apollo 8's orbital earthrise over the Moon, Buzz Aldrin triumphantly facing Armstrong's lens on Apollo 11, the whole Earth of Apollo 17 floating so delicately alone in the indifferent velvet of the void – and they deserve to be, having completely transformed the way humans conceive of themselves, but they also suffer from over-exposure. I had chanced across a few of the astronauts' more obscure black-and-white images in a small exhibition catalog, however, and their spare beauty and quality of light showed a different Moon than the one we had all collectively come to know, a freshly compelling one. I had a hunch there was more, and in 1994 resolved to visit NASA and discover just how much. What I found in a small, windowless concrete bunker that autumn in Houston, Texas, came to occupy me for the next four years and has resulted in this book.

The Apollo inflight photographic archive is extraordinary. First, it's vast: the astronauts exposed miles of 70mm hand-held Hasselblad black-and-white negatives and color transparencies, as well as monitored giant automatic mapping cameras on the later missions that shot 5" x 5" and panoramic images of the lunar surface from orbit. Altogether more than 17,000 hand-held and 15,000 automatic orbital photographic images were made. Secondly, the archive has depth as well as breadth: the space explorers photographed just about everything, all the time, and the different lunar landscapes that each successive mission visited were varied and hauntingly beautiful. Finally, the overall quality and technical proficiency of the images are stunning. The pictures are so sharp and crisp they verge on the surreal; that they were made in a vacuum certainly helps, but there is no question that the astronauts knew how to operate their cameras superbly.

\+ + +

A NOTE ON THE PHOTOGRAPHS

Full Moon has used the latest in digital techniques to obtain the highest-quality photographic reproduction possible, resulting in a major advance over past NASA procedures for providing imagery to the public. When the Apollo missions returned to Earth, the Agency duplicated the fragile and precious original film once before putting it into cold storage. These "master dupes" were then used to make succeeding copies when an image request was received. In the past, the masters were never allowed out of NASA's possession, so publishers and exhibitors have had to make do with fourth- and fifth-generation duplicates, resulting in image deterioration and substantial loss of information with each succeeding generation.

By negotiating permission from NASA to take the masters offsite and digitally scan them at film resolution, *Full Moon* has circumvented this unfortunate procedure. The advent of digital imaging has made it possible to "clone" rather than "reproduce" an image, avoiding the addition of successive generations and thus keeping far more visual information intact. Application of digital technology at the source also allows subtle control of color balance, contrast, and density in ways that far surpass conventional analog darkroom techniques, as well as the ability to delicately composite separate images into panoramas.

Because of the scientific- and documentary-survey nature of the Apollo images, I have been especially careful in my use of powerful digital tools not to alter them beyond what any good printer might do in making a fine exhibition print. Naturally, even this required a host of aesthetic decisions for each image – for which I am solely responsible – but in each case I let the information on the film lead me, not vice versa. Color was the most demanding challenge. As explained in the essays in this book, the physics of human color-perception on the Moon are more complex and subjective than on Earth. Film itself adds its own peculiar characteristics to a recorded image, and the type of film and processing methods used also varied from mission to mission. Color accuracy is further complicated by the fact that in the process of duplication, NASA's masters often gained a blue, green, cyan, or yellow cast that simply is not present on the original film. My guide has been the fact that lunar soil, when held in the hand on the Moon's surface, appears as lighter and darker shades of gray.

In general, all images appear in the book as they were made on the missions. All black-and-white reproductions were shot on black-and-white negative film, all color reproductions on color transparency film. With the exception of certain double-page spreads and Earth images floating in black space, I have kept image cropping to a minimum, holding to full-frame reproduction as a matter of course. I have taken liberties with four images, however, in each case digitally removing certain small items for aesthetic reasons. In images 26 and 27, I have removed a thin sensor probe that intruded into the picture frame from the orbiting command module. In images 72 and 78, I have removed the reseau lens grid marks. Finally, I have added a slight, digitally induced blur to the enlargement of image 58, the family photograph on the surface of the Moon.

Regarding captions, I have checked and rechecked each to ensure the highest accuracy. Errors may remain, however, and I apologize for them in advance. Accurate attribution of inflight imagery to the particular astronaut who shot it is more difficult than with surface material because there are fewer hard records from which to reconstruct the sequence of events. In cases where doubts remain, I have given a "probably by" attribution to the mission's command module pilot, who was in general formally assigned most of a mission's inflight photographic duties by NASA.

For information on digitally-scanned and processed Apollo imagery, contact *www.projectfullmoon.com*, or call 1-800-370-0093, San Francisco, USA.

Jacket: Leaving the scene of their explorations, the Apollo 15 astronauts photograph the Moon's southern hemisphere shortly after firing their rocket engine to begin the trip back to Earth. The image includes a portion of the lunar farside, which is not visible from Earth. This view was captured by a high-resolution camera stored in the side of the service module. Metric mapping camera black-and-white negative by Alfred Worden, Apollo 15, July 26–August 7, 1971.

Boards: The battered highlands of the lunar farside, seen from some 750 miles at the start of Apollo 16's return to Earth. Visible on the front cover are craters Saha at center top and Moiseev at center bottom, while 47-mile-wide Crater King with its distinctive "lobster claw" interior appears to the far left on the rear cover, with a portion of Crater Lobachevsky at the lower right. Each cover shows an approximately 200-mile-wide square section. Metric mapping camera black-and-white negative by Kenneth Mattingly, Apollo 16, April 16–27, 1972.

Billowing exhaust smoke, the 5 massive first-stage engines of the Apollo 11 Saturn rocket ignite to begin the 238,000-mile journey that would land Neil Armstrong and Buzz Aldrin on the surface of the Moon days later. One of the rocket's 4 fins, labeled with the letter "B" for clear identification, appears through the exhaust gases. NASA ran between 20 and 30 remote-controlled and fire-insulated motion-picture cameras close to the engines during each Apollo launch, to have a record for later analysis in case of any malfunction. Fortunately, this safety footage never had to be used for its intended purpose on any of the Apollo missions. Remote-controlled motion-picture camera film still, Apollo 11, July 16, 1969.

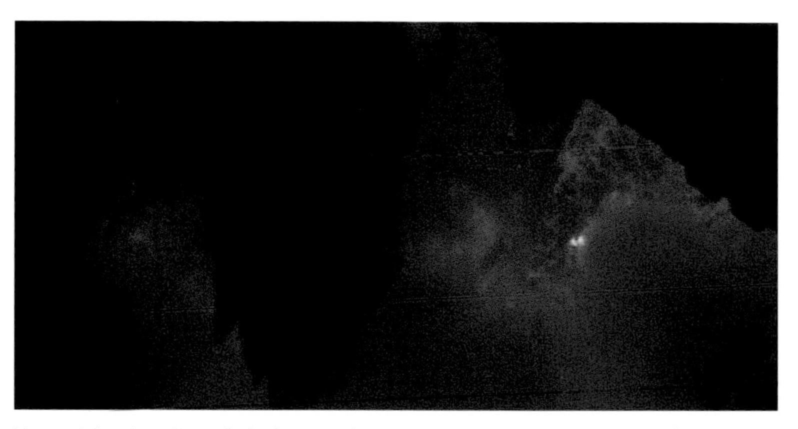

Mounted directly underneath the Saturn rocket, a camera captures the exact instant of ignition of one of the rocket's 5 F-1 engines. Remote-controlled motion-picture camera film still, Apollo 11, July 16, 1969.

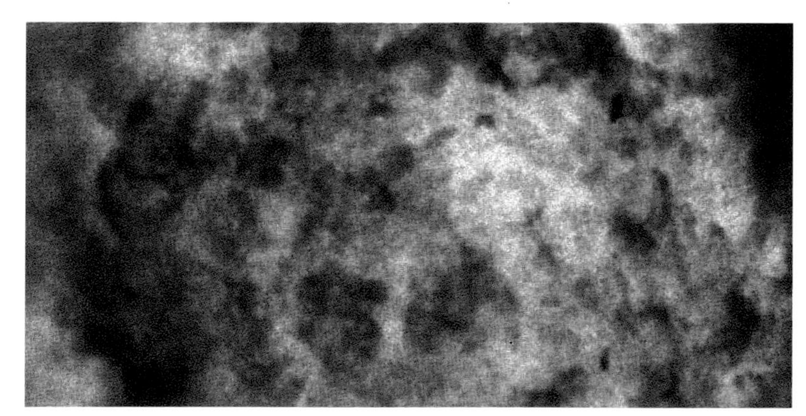

An instant later, the camera captures the growing yellow fireball of kerosene and pure oxygen that will soon turn into a white-hot wall of flame exceeded in power only by a nuclear explosion. Cameras, equipment, and the launch pad itself were doused with a constant spray of water during Apollo launches, to prevent damage from the searing heat and blast of the rocket's exhaust and also to suppress the deafening roar. Remote-controlled motion-picture camera film still, Apollo 11, July 16, 1969.

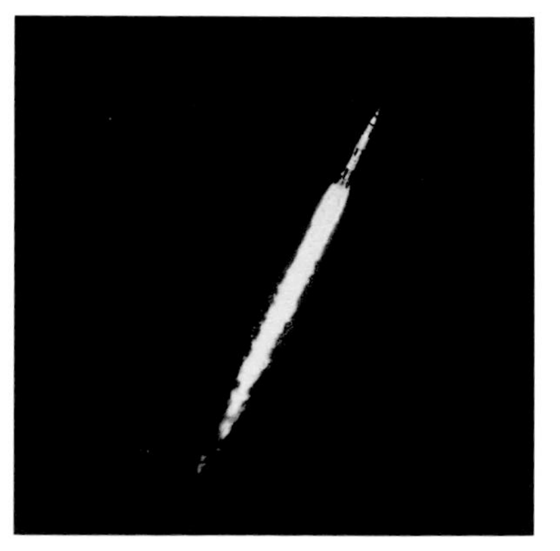

1. The first test flight of the Apollo Saturn V Moon rocket. At liftoff the 363-foot vehicle weighed 6,220,025 pounds, with the first-stage engines producing 7,500,000 pounds of thrust and burning 15 tons of fuel per second. Color transparency taken by the Intermediate Ground Optical Recording Camera (IGOR) shortly after liftoff, Apollo 4 (unmanned), November 9, 1967.

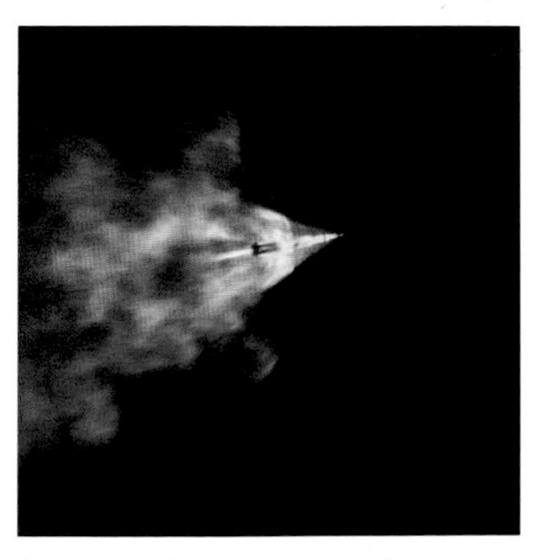

2. About 41 miles above Earth the first stage of the Saturn V rocket is depleted; it separates with a plume of smoke and flame. An instant later, the 5 engines of the second stage will ignite and burn for another 6.5 minutes, boosting the rocket to an altitude of 116 miles. Color transparency made by a 70mm tracking camera on a U.S. Air Force chase plane flying at 40,000 feet, Apollo 11, July 16, 1969.

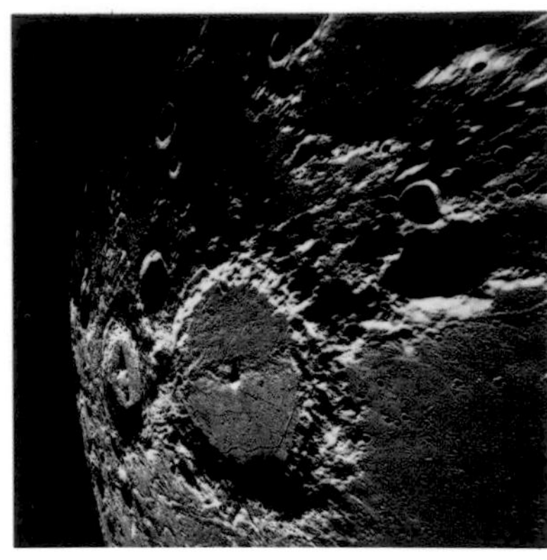

26. Visible are craters Arzachel at the left, Alphonsus to the center, and Ptolemaeus to the right in this orbital view made at an altitude of 68 miles. Alphonsus is 67 miles wide. Metric mapping camera black-and-white negative made by Kenneth Mattingly, Apollo 16, April 16–27, 1972.

27. A forbidding landscape created by debris that was ejected during the formation of the giant Imbrium impact basin. This so-called Imbrium sculpture affects large portions of the lunar near-side. Ridges and grooves point toward the center of the Imbrium basin, which is several hundred miles away. Metric mapping camera black-and-white negative at an altitude of 68 miles, by Kenneth Mattingly, Apollo 16, April 16–27, 1972.

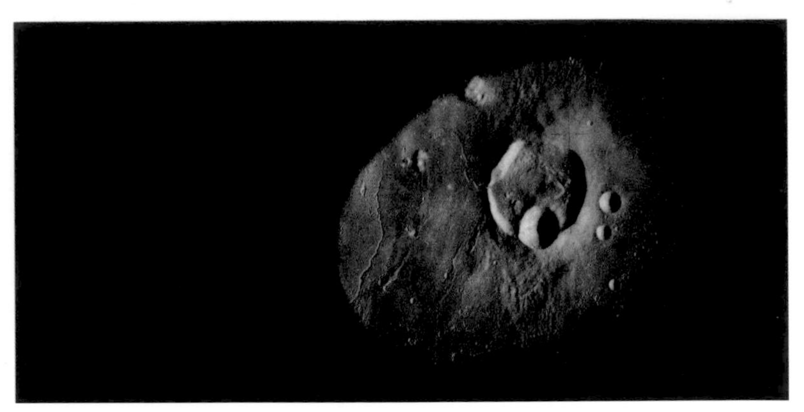

28. Crater Krieger, 12 miles in diameter, seen through one of the windows of the command module *Endeavor* from an orbital altitude of 66 miles (see caption 128). Hasselblad 70mm black-and-white negative by Alfred Worden, Apollo 15, July 26–August 7, 1971.

29. An enlarged detail of the battered highlands of the lunar farside, seen from some 1,000 miles at the start of Apollo 16's return to Earth. The 47-mile-wide Crater King, with its distinctive center in the shape of a lobster claw, can be seen to the left. Metric mapping camera black-and-white negative by Kenneth Mattingly, Apollo 16, April 16–27, 1972.

30. Earthrise seen for the first time by human eyes. Apollo 8 astronaut William Anders took this photograph, which looks southwest toward Crater Gibbs from an orbital altitude of about 70 miles. For years after the mission, commander Frank Borman claimed to have made the picture, but onboard voice recordings reveal that this photo, as well as the more famous color views taken immediately after, were made by Anders. Hasselblad 70mm black-and-white negative, Apollo 8, December 24, 1968.

31. Battered highlands on the lunar farside, looking south in a view covering an area about the size of Switzerland. Hasselblad 70mm black-and-white negative probably by Michael Collins, Apollo 11, July 16–24, 1969.

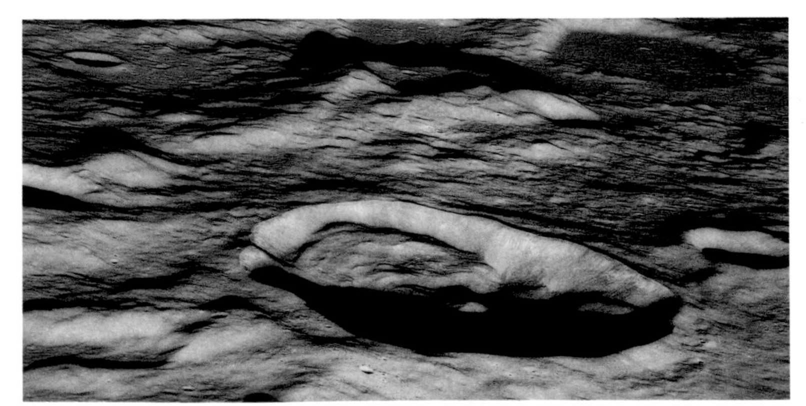

32. An enlarged detail of the crater Godin, about 27 miles wide, seen from an altitude of 69 miles and located in the highland region that separates the Sea of Tranquillity from the Central Bay. Hasselblad 70mm black-and-white negative probably by John Young, Apollo 10, May 18–26, 1969.

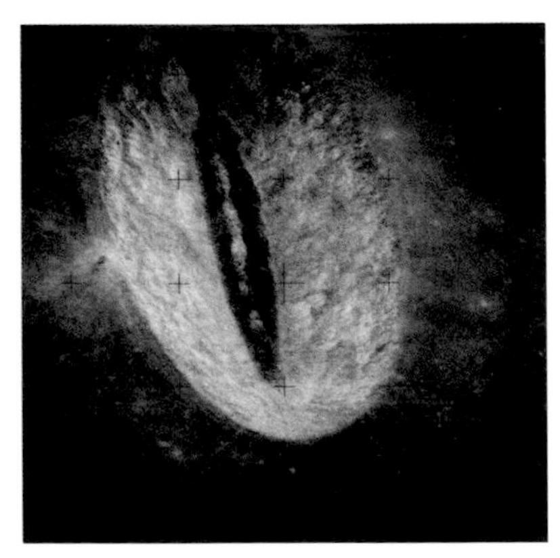

33. Crater Messier in the Sea of Fertility, seen from an altitude of 70 miles with a 500mm telephoto lens. Messier is believed to have been formed by a low-angle meteoric impact, creating a furrow 5 miles wide and 9 miles long. Hasselblad 70mm black-and-white negative probably by Alfred Worden, Apollo 15, July 26–August 7, 1971.

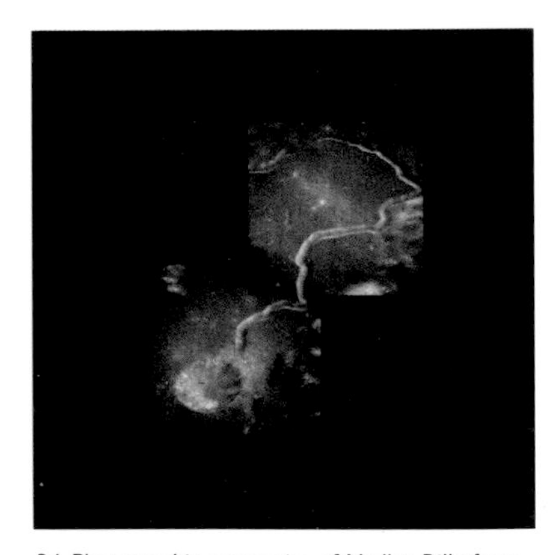

34. Photographic composite of Hadley Rille from an orbital altitude of 70 miles, showing the Apollo 15 landing site in the lower right of the upper photograph, just below the rille. The Apennine mountains are visible to the right in the upper photograph, and show about 15,000 feet of relief; Hadley Rille itself is almost a mile wide and 1,000 feet deep near the landing site (see captions 72 and 73). Hasselblad 70mm black-and-white negatives, by Alfred Worden, Apollo 15, July 26–August 7, 1971.

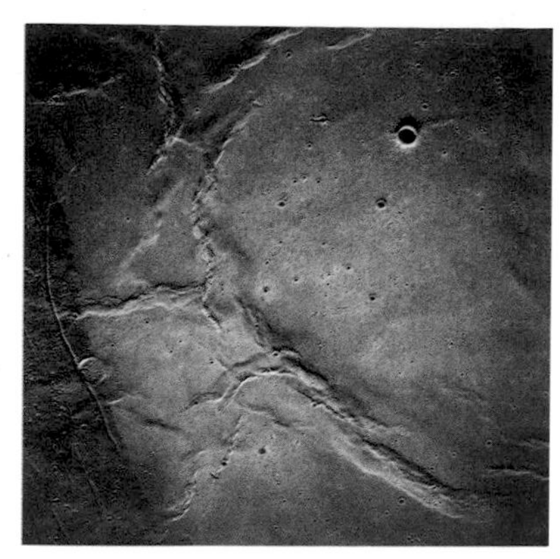

35. Crater Condorcet, west of the Sea of Crises, photographed from an orbit 75 miles above the lunar surface. The low ridges, called wrinkle ridges, were formed as the vast lava plain cooled. Condorcet is 46 miles wide. Metric mapping camera black-and-white negative by Ronald Evans, Apollo 17, December 7–19, 1972.

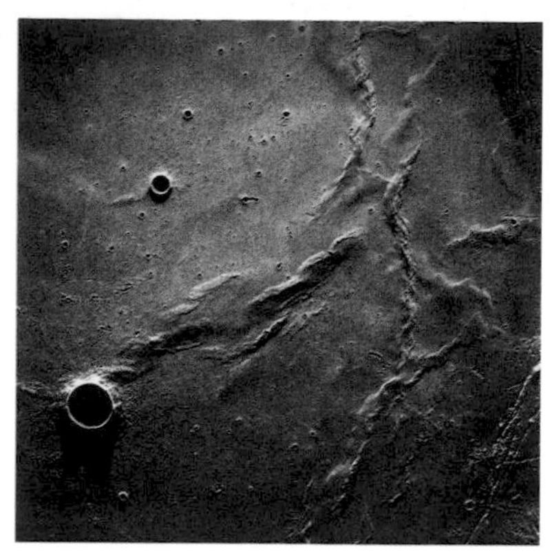

36. Crater Condorcet, with Bessel below, photographed 69 miles above the lunar surface in the Sea of Serenity. The area shown is about 105 square miles. Metric mapping camera black-and-white negative by Ronald Evans, Apollo 17, December 7–19, 1972.

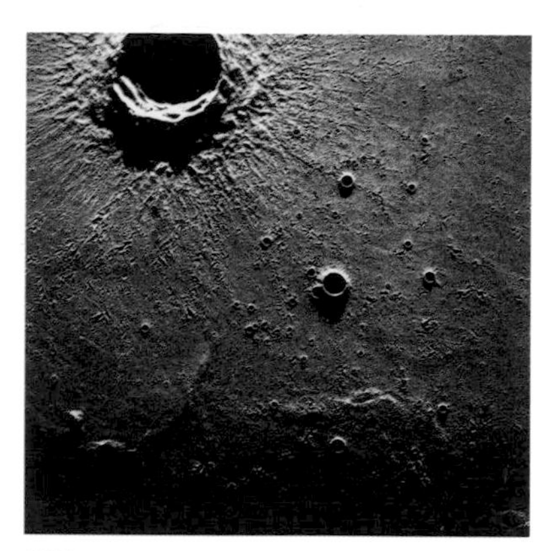

37. The metric mapping camera, operating automatically, took this sequence of pictures as the command module *Endeavor* drifted over the nearside crater Timocharis 62 miles below. This view is one of hundreds obtained by command module pilot Alfred Worden while his crewmates explored the lunar surface. Timocharis is 20 miles wide. Metric mapping camera black-and-white negatives, Apollo 15, July 26–August 7, 1971.

38. Timocharis Crater, Apollo 15.

39. Timocharis Crater, Apollo 15.

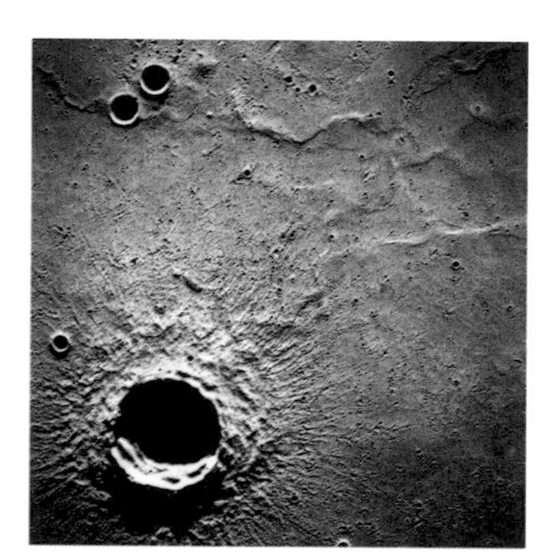

40. Timocharis Crater, Apollo 15.

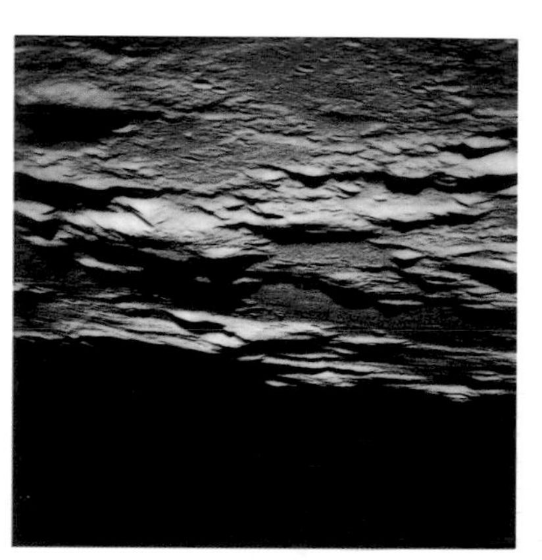

41. Floodlit by the morning Sun, the lunar highlands to the west of the Sea of Tranquillity stand out in stark relief. The twin craters Sabine and Ritter can be seen to the upper left, with 7-mile-wide crater Schmidt to the center of the image. Hasselblad 70mm black-and-white negative made from an altitude of about 70 miles, probably by Michael Collins, Apollo 11, July 16–24, 1969.

42. The command module *America* photographed from the ascent stage of the lunar module *Challenger* just before docking and the return of the moonwalking astronauts to their orbiting colleague. The Sea of Fertility is visible below, as the dark surface area on the upper-right horizon. Hasselblad 70mm transparency by Harrison Schmitt, Apollo 17, December 7–19, 1972.

43. Readying for final descent, the lunar module *Intrepid* floats 69 miles above the giant crater Ptolemaeus in this westward-looking view, after having separated from the command module *Yankee Clipper*. The circular crater in the middle distance on the right is Herschel. Touchdown for *Intrepid* will be on the Ocean of Storms. Hasselblad 70mm transparency by Richard Gordon, Apollo 12, November 14–24, 1969.

44. A view of lunar module *Intrepid's* shadow from the commander's window, just after landing and prior to surface excursions on the Ocean of Storms. The lander is 23 feet tall. Hasselblad 70mm black-and-white negative by Alan Bean, Apollo 12, November 14–24, 1969.

45. Astronaut's shadow. Hasselblad 70mm black-and-white negative by Harrison Schmitt, Apollo 17, December 7–19, 1972.

46. Unfiltered by any atmosphere, the Sun as seen from the lunar surface is more brilliant than on Earth. Lens flare creates the bluish cast seen in this photograph. The Apollo landings took place during the early lunar morning, with the Sun about 12 degrees above the horizon, to aid the astronauts to spot craters and boulders. Hasselblad 70mm transparency near Surveyor Crater by Charles Conrad, Apollo 12, November 14–24, 1969.

47. Apollo 11 moonwalker Edwin "Buzz" Aldrin took this unusual close-up photograph of pristine lunar soil by hand-holding the camera above the ground. Moments later, he would disturb this aeons-old patch of ground with his own boot. Hasselblad 70mm transparency, Apollo 11, July 16–24, 1969.

48. Aldrin's boot print, in an image that has come to symbolize human exploration of the Moon. The fine dust clearly records the pattern of treads on the sole of the boot. Scientists estimate that the constant rain of micrometeorites onto the Moon is sufficient to churn the uppermost half-inch of lunar soil every 10 million years. Accordingly, this footprint should last one or two million years. Hasselblad 70mm transparency by Edwin Aldrin, Apollo 11, July 16–24, 1969.

49. Alan Bean at Sharp Crater with the handtool carrier near his right hand. On the carrier are visible the cuplike sample bags and the large sample collection bag that fills its center. Hasselblad 70mm black-and-white negative by Charles Conrad, Apollo 12, November 14–24, 1969.

50. Edwin Aldrin carrying a package of scientific experiments to a deployment site south of the lunar module *Eagle*. In his left hand is a seismometer (see caption 57) used to test hypotheses about the inner composition of the Moon, and in his right, a laser-ranging retroreflector experiment used to measure its exact distance from the Earth. Hasselblad 70mm transparency by Neil Armstrong, Apollo 11, July 16–24, 1969.

51. The lunar module *Antares* at Fra Mauro, with the intense Sun just behind. Astronauts Alan Shepard and Edgar Mitchell fondly referred to their lunar home and oasis as "jewel-like." Hasselblad 70mm transparency by Alan Shepard, Apollo 14, January 31–February 9, 1971.

52. The descent engine bell and leg of lunar module *Intrepid*, showing the cylindrical tube of the nuclear fuel cask to the upper right; its fuel element has just been placed in the radioisotope thermoelectric generator (RTG) (see caption 56), a small nuclear power station that will power several lunar surface experiments for years to come. Hasselblad 70mm black-and-white negative by Alan Bean, Apollo 12, November 14–24, 1969.

53. Using an experiment nicknamed "the thumper," Edgar Mitchell detonates a small explosive charge to create a miniature "moonquake." Wires leading from the thumper carry sensors to measure the vibrations produced, which reveal information about the structure of the lunar surface. In the background is a crater the astronauts have christened "Old Nameless." Hasselblad 70mm transparency by Alan Shepard, Apollo 14, January 31–February 9, 1971.

54. Because the Moon has no atmosphere, it is an ideal place to study the solar wind, the stream of subatomic particles emitted by the Sun. Pictured is the solar wind composition experiment (SWC), a strip of foil measuring 4.5 feet by 1 foot. Each square centimeter was peppered with 6 to 7 million solar wind particles each second, and was left out for 18 hours before being packed up for return to scientists on Earth. Hasselblad 70mm black-and white-negative by Alan Bean, Apollo 12, November 14–24, 1969.

55. Alan Bean deploying experiments on the lunar surface; the blue aura that surrounds him is thought to be water-vapor ice crystals emitted from the boiler on his space-suit backpack, though neither Bean nor his commander reported seeing such vapor from the backpacks on the surface. Hasselblad 70mm transparency by Charles Conrad, Apollo 12, November 14–24, 1969.

56. On 5 of the 6 missions, the astronauts deployed the Apollo lunar surface experiments package (ALSEP). The ALSEP's central station, which transmits data to Earth, is at rear, outfitted with gold Mylar insulation. In front sits the radioisotope thermoelectric generator (RTG). Inside, a plutonium-238 fuel capsule (see caption 52) generates a temperature of 1,350 degrees F. Thermocouples convert this energy into power for the experiments. Hasselblad 70mm transparency by Alan Shepard, Apollo 14, January 31–February 9, 1971.

57. ALSEP's lunar seismometer, officially designated the passive seismic experiment (PSE), deployed at the Hadley Plains. The seismometer is surrounded by an apron of aluminized Mylar to reduce fluctuations in surface temperature. Recording moonquakes and meteorite impacts, the seismometer enabled scientists to draw profiles of the lunar interior. Hasselblad 70mm photograph by David Scott, Apollo 15, July 26–August 7, 1971.

58. A snapshot of Charles Duke and his family in their Houston, Texas, backyard, left at the Moon's Descartes Highlands. Hasselblad 70mm transparency by Charles Duke, Apollo 16, April 16–27, 1972.

59. Footprints on the interior slope of Surveyor Crater. In the climax of the Apollo 12 mission, astronauts Charles Conrad and Alan Bean descended into the 656-foot wide crater to visit the unmanned Surveyor III probe, which had landed on the Moon 2½ years earlier. In history's first pinpoint lunar landing, the astronauts touched down only 600 feet from the probe. Hasselblad 70mm black-and-white negative by Alan Bean, Apollo 12, November 14–24, 1969.

60. Having abandoned their hunt for the rim of Cone Crater, astronauts Alan Shepard and Edgar Mitchell visited this boulder field, which shows a 3-foot-high boulder at left. The astronauts did not realize that at this moment they were looking past the rim itself, which was less than 100 feet away. Hasselblad 70mm black-and-white negative by Edgar Mitchell, Apollo 14, January 31–February 9, 1971.

61. A shadow portrait of the Apollo 12 astronauts as they prepared to collect rock and soil samples. At right sits a device called a gnomon, used to provide a photographic scale by which to judge sizes of features. The gnomon indicated local slope and, by its shadow, the Sun angle; it also included a color bar to help analysts match photographic colors and tones to those of the actual scene. Hasselblad 70mm black-and-white negative by Alan Bean, Apollo 12, November 14–24, 1969.

62. A view of a color and contrast photographic chart which the astronaut has positioned on the sunlit, western wall of a small crater to use as a control photograph. Hasselblad 70mm black-and-white negative by Alan Bean, Apollo 12, November 14–24, 1969.

63. "Contact Rock" at the rim of Cone Crater. In the distance sits the lunar module *Antares*, a tiny speck in the Fra Mauro Highlands. (To find the craft, look at the uppermost row of reseau calibration marks. *Antares* sits to the upper left of the second mark from the photo's right-hand edge.) Hasselblad 70mm black-and-white negative, by Edgar Mitchell, Apollo 14, January 31–February 9, 1971.

64. Commander John Young and the battery-powered lunar rover in a photographic composite of the Descartes Highlands, showing the lunar module *Orion* to the left and Stone Mountain, some 1,600 feet high, to the right. The rover, present on the last three Apollo missions, was able to carry two suited astronauts, their gear and cameras, and several hundred pounds of bagged lunar samples, for miles across the lunar surface. Hasselblad 70mm transparencies by Charles Duke, Apollo 16, April 16–27, 1972.

65. Tracks from the modular equipment transporter (MET) gleam in the harsh lunar sunlight. This view looks east toward the lunar module *Antares*, roughly 600 feet away. Loaded with tools, rock samples, and cameras, the MET would have weighed 150 pounds on Earth but weighed only 25 in the Moon's one-sixth gravity. As a result, its tracks are only ¾ of an inch deep. Hasselblad 70mm transparency by Alan Shepard, Apollo 14, January 31–February 9, 1971.

66. A glance backward on the way to the top of Cone Crater during the second moonwalk of Apollo 14; the edge of the MET is visible at left, its tracks curving out of the frame. Hasselblad 70mm black-and-white negative by Alan Shepard, Apollo 14, January 31–February 9, 1971.

67. Charles Duke seen twice in a photographic composite of Apollo 16's first moonwalk, next to 120-foot-wide Plum Crater in the Descartes Highlands. Stone Mountain, 1,600 feet high and 3 miles distant, forms the skyline to the right. On the left, Duke bores into the lunar soil to extract a core sample; on the right, he moves on to the next task. Hasselblad 70mm transparencies by John Young, Apollo 16, April 16–27, 1972.

68. The southern face of "Saddle Rock"; part of a set of large outcroppings just 75 feet shy of the rim of Cone Crater. This image shows the astronauts' geologic hammer and cuplike sample collection container. Scientists believe these rocks were blasted out of the Moon's crust by the gigantic impact that formed the Imbrium Basin. Hasselblad 70mm black-and-white negative by Edgar Mitchell, Apollo 14, January 31–February 9, 1971.

69. Astronaut Alan Bean holds a special sample container designed to aid studies of the lunar environment. He and Charles Conrad have just filled the container with lunar soil at Sharp Crater on the Ocean of Storms. Bean's checklist for the moonwalk is visible on his left wrist, along with his Hasselblad camera mounted on the control unit on his chest. His gold-plated sun visor clearly reflects Conrad, the photographer. Hasselblad 70mm black-and-white negative, Apollo 12, November 14–24, 1969.

70. Commander David Scott test-drives the first manned lunar rover on Apollo 15's first surface excursion; note the orbital lunar photographs used as an aid to navigation. The rover weighed 462 pounds on Earth but only 77 on the Moon, and accordingly gave quite a spirited ride. Seatbelts were mandatory. Hasselblad 70mm black-and-white negative by James Irwin, Apollo 15, July 26–August 7, 1971.

71. Photographic composite of rover tracks looking west on a steep 17-degree grade of Hadley Delta near Spur Crater. At left is the slope that rises to the peak of the mountain, more than 11,000 feet above the Marsh of Decay below. Though the rover was highly mobile, the soft soil and precipitous grade here caused it to begin sliding down the hill when parked. In general, crossing a slope was nerve-racking for the astronaut on the downhill side, but there were no rollovers in any of the missions equipped with a rover. Hasselblad 70mm black-and-white negatives by James Irwin, Apollo 15, July 26–August 7, 1971.

72. Hadley Rille, seen looking northwest from the slopes of Hadley Delta mountain near St. George Crater. Measuring 1 mile across and 1,000 feet deep, the winding canyon is littered with boulders and continues for almost 80 miles along the edge of the Marsh of Decay. Rilles like Hadley may have been formed originally as underground lava tubes, only to have their roofs later bombarded away by aeons of meteoric impact. Hasselblad 70mm black-and-white negative by James Irwin, Apollo 15, July 26–August 7, 1971.

73. The bottom of Hadley Rille, seen looking north from St. George Crater. The clarity of the lunar vacuum and the foreshortening of the telephoto lens offer a deceptively small sense of scale; actually, the largest boulder at the bottom is 45 feet wide. A feature called the Terrace, located on the rille's eastern wall, is visible in the background. Hasselblad 70mm black-and-white negative made with a 500mm lens by David Scott, Apollo 15, July 26–August 7, 1971.

74. A crater, fault scarp, and 7,000-foot mountain are visible in, respectively, the fore, middle, and backgrounds of this image taken at geology station 3 on the way to Shorty Crater from Ballet Crater. This part of the Taurus-Littrow valley was buried when projectiles from the Tycho impact, 1,400 miles to the southwest, started an avalanche. Hasselblad 70mm black-and-white negative by Harrison Schmitt, Apollo 17, December 7–19, 1972.

75. A view of one of the most valuable tools used by the moonwalkers: a set of long-handled tongs, used to pick up rock samples. The tongs were opened and closed by means of a T-shaped handle at the upper end. Space suits worn by the astronauts on the first three lunar landings were very stiff at the waist. Even on the last three landings, which featured more flexible suits, the tongs were essential. Hasselblad 70mm black-and-white negative made near Scarp Crater by David Scott, Apollo 15, July 26–August 7, 1971.

76. At Spur Crater, on the slopes of Hadley Delta, Dave Scott and Jim Irwin discovered a sample of the Moon's primordial crust. Dubbed the "Genesis Rock" by journalists, its age has been measured at 4.5 billion years, only 100 million years younger than the solar system itself. On their discovery, an elated Scott declared, "We've found what we came for." In this view, Scott manipulates collection tongs at Spur. Hasselblad 70 mm black-and-white negative by James Irwin, Apollo 15, July 26–August 7, 1971.

77. Broken layers of basalt, once deposited as lava flows, poke out of the dust-covered wall of Hadley Rille. Although geologists had obtained extensive evidence that the lunar seas were composed of layered volcanic flows, Hadley Rille provided the only example of in-situ rock layers visited on the lunar surface. Hasselblad 70mm black-and-white negative made with a 500mm lens by David Scott, Apollo 15, July 26–August 7, 1971.

78. The tiny lunar module *Challenger* seen from almost 2 miles away, photographed with a 500mm lens from the base of the North Massif. The large blocky craters to the right are Camelot and Horatio, while the sides of the South Massif rise steeply 5 miles away in the background. Hasselblad 70 mm black-and-white negative by Eugene Cernan, Apollo 17, December 7–19, 1972.

79. Image composite of mission commander David Scott photographing a geologic find before bagging it, 300 feet up the flank of 11,500-foot-high Hadley Delta mountain. The white spots above him are lens flares caused by shooting directly east into the Sun; the mountains behind him are 10½ miles away. Hasselblad 70mm black-and-white negatives by James Irwin, Apollo 15, July 26–August 7, 1971.

80. A boulder field near the eastern rim of Hadley Rille, with Hadley Delta mountain in the background. Hasselblad 70mm black-and-white negative by James Irwin, Apollo 15, July 26–August 7, 1971.

81. On the slopes of the North Massif, Eugene Cernan and Harrison Schmitt visited a house-sized, five-piece boulder known as "Split Rock" (see captions 84 and 86). The boulder, a portion of which is seen here, is composed of material ejected from the giant impact basin located at the Sea of Serenity. Cernan's arm and leg are seen at right; the South Massif is visible at upper left, beyond the boulder. Hasselblad 70mm black-and-white negative by Harrison Schmitt, Apollo 17, December 7–19, 1972.

82. Geologist-astronaut Harrison Schmitt, the first professional scientist to walk on another world, stands next to the rover at the rim of 360-foot-wide Shorty Crater. It was at Shorty that Schmitt discovered an unusual orange soil, which is visible in subtle patches near the rover and in streaks on the crater walls. At first, Schmitt believed the orange coloration was the result of volcanic gases escaping through cracks in the lunar crust in relatively recent lunar geologic history. However, samples of the rare soil analyzed on Earth proved to be 3.7 billion years old. They consist of microscopic beads of glass, colored by titanium, created by volcanic "fire fountains" that sprayed forth from deep within the Moon's interior. Behind Schmitt, on the horizon almost 4 miles away, Family Mountain rises among the peaks of the Taurus-Littrow mountains. Hasselblad 70mm transparencies by Eugene Cernan, Apollo 17, December 7–19, 1972.

83. Alan Bean placing a double-core tube in the ground at Halo Crater, about to push it in as far as he can before hammering it to full depth. Core samples of the lunar soil were an invaluable method for deciphering its history. Hasselblad 70mm black-and-white negative by Charles Conrad, Apollo 12, November 14–24, 1969.

84. Photographic composite of Eugene Cernan and the rover near Split Rock (see captions 81 and 86). The boulder – now broken into five separate pieces – is a breccia, or rock made of fragments of other rocks, that has rolled down the side of the steep North Massif in a 1,500-foot-long furrow. The Sculptured Hills are visible beyond the boulder, and the South Massif can be seen to the far right. Cernan leans on an instrument called a gravimeter, used to measure variations in the lunar gravity field. Hasselblad 70mm black-and-white negatives by Harrison Schmitt, Apollo 17, December 7–19, 1972.

85. Charles Duke has planted the long-handled scoop in the dust while John Young, at left, stands by holding the geology hammer. Beyond sits the lunar rover. The shadow of Duke's helmet intrudes at lower right to create a simple but beautiful visual composition. Hasselblad 70mm black-and-white negative by Charles Duke, Apollo 16, April 16–27, 1972.

106. In NASA's dress rehearsal for the lunar landing, the Apollo 10 command module *Charlie Brown* prepares to dock with the lunar module *Snoopy* in lunar orbit. The command module is coated with reflective material to mirror the Sun's heat. About 12 feet in diameter and weighing 12,500 pounds, it had about as much habitable volume as a walk-in closet (see caption 5). Hasselblad 70mm transparency by Eugene Cernan, Apollo 10, May 18–26, 1969.

107. A slender crescent Earth rises over the lunar highlands in this composite of two images made at an altitude of 70 miles. The steep slopes of the southwestern inner rim of Humboldt Crater are visible towards the left edge of the horizon. Hasselblad 70mm transparency by Alfred Worden, Apollo 15, July 26–August 7, 1971.

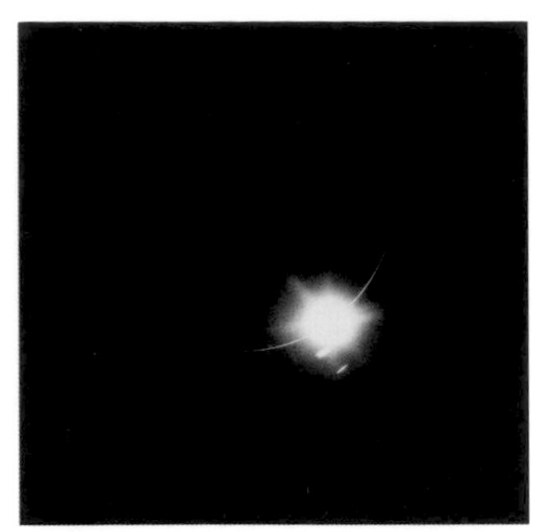

108. During their trip home, the Apollo 12 astronauts became the first people to witness an eclipse of the Sun by the Earth, here photographed shortly before atmospheric reentry. Hasselblad 70mm black-and-white negative probably by Richard Gordon, Apollo 12, November 14–24, 1969.

109. A view of the Moon never seen before the space age, captured at about 1,000 miles as the astronauts began their trip back to Earth. The image is centered on the boundary between the lunar near side and the Moon's hidden face. Three lunar seas are visible as dark patches, clockwise from upper left: the Sea of Crises, the Border Sea, and Smyth's Sea. At lower right are the highlands of the lunar farside. Metric mapping camera black-and-white negative by Kenneth Mattingly, Apollo 16, April 16–27 1972.

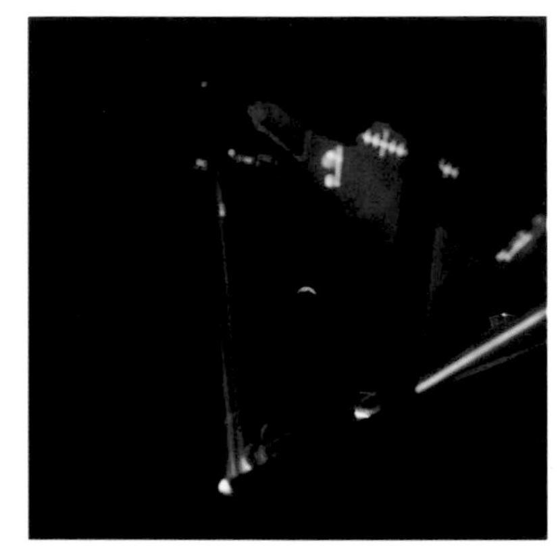

110. A distant Earth seen through the window of the lunar module *Aquarius* during Apollo 13's long and arduous return voyage. After the explosion of an oxygen tank rendered the command module *Odyssey* useless, the astronauts used *Aquarius* as a lifeboat. The astronauts spent days in the darkened cabin, with all non-essential systems shut down to save electrical power (see caption 20). Hasselblad 70mm transparency probably by John Swigert, Apollo 13, April 11–17, 1970.

111. The shiny metal skin of the Apollo 10 lunar module *Snoopy* is seen from the command module *Charlie Brown*, which is docked to *Snoopy's* roof. At center is the lunar module's overhead window, used by the commander during rendezvous maneuvers. The grid marks on the window are used by the commander to judge the relative orientation of the command module and lunar module as he closes in. Hasselblad 70mm transparency by John Young, Apollo 10, May 18–26, 1969.

112. Earth, seen beyond the lunar module *Eagle*, photographed from the command module *Columbia* as the joined craft sped moonward. A set of the lunar module's maneuvering thrusters is seen at top; part of the craft's circular docking target is visible at bottom. Hasselblad 70mm transparency probably by Michael Collins, Apollo 11, July 16–24, 1969.

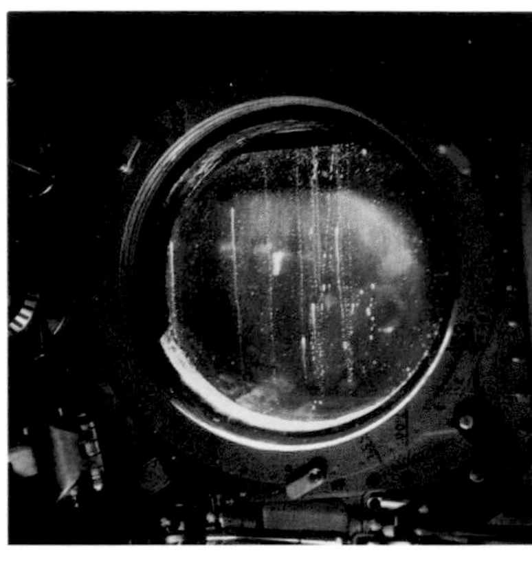

113. The hatch window of command module *Yankee Clipper*; the condensation inside the glass window was due to a design flaw that was corrected on later flights. Hasselblad 70mm transparency by Richard Gordon, Apollo 12, November 14–24, 1969.

114. For the first time on any Apollo flight, the Apollo 17 astronauts viewed a full Earth as they began their lunar journey. Africa and the Arabian peninsula are clearly visible. The African rain forests are hidden by clouds; the Sahara Desert to the north and Kalahari to the south are clear. It is late spring in the southern hemisphere, and the Antarctic icecap is seen in full sunlight. Hasselblad 70mm transparency by Harrison Schmitt, Apollo 17, December 7–19, 1972.

115. Walter Cunningham sleeps inside the command module in Earth orbit. Hasselblad 70mm transparency by Walter Schirra, Apollo 7, October 11–22, 1968.

116. A half-Earth gleams in the blackness of space as Apollo 13 heads for the Moon. Clouds hide much of North America, but the southwestern United States and part of Mexico are clearly visible. Hasselblad 70mm transparency probably by John Swigert, Apollo 13, April 11–17, 1970.

117. The Apollo 16 astronauts enjoyed this spectacular view of "the blue planet" shortly after leaving Earth orbit. Much of the United States is visible; note the contrast between the arid lands of the American west and the greener regions to the east. The solid white area to the upper left is the Arctic ice cap, with snow-covered terrain below in Canada. Hasselblad 70mm transparency probably by Kenneth Mattingly, Apollo 16, April 16–27, 1972.

118. On board the unmanned Apollo 4 command module, an automatic camera snapped this view of the crescent Earth looking west over the Atlantic from a distance of about 10,000 miles. Apollo 4, the first test flight of the Saturn V Moon rocket, also tested the command module's heatshield in a high-speed reentry similar to those that would be experienced on lunar missions. Automatic Hasselblad transparency, Apollo 4 (unmanned), November 9, 1967.

119. The orange hues of the Horn of Africa contrast with the deep blues of the Earth's oceans in this view made shortly after the translunar injection maneuver. Hasselblad 70mm transparency by Harrison Schmitt, Apollo 17, December 7–19, 1972.

120. Mexico fills this view by the Moon-bound astronauts, showing the eastern and western Sierra Madre mountain ranges. Hasselblad 70mm transparency by Michael Collins, Apollo 11, July 16–24, 1969.

121. The far limb of Earth photographed just before atmospheric reentry, showing late afternoon, sunset and night over Africa. Hasselblad 70mm transparency by Michael Collins, Apollo 11, July 16–24, 1969.

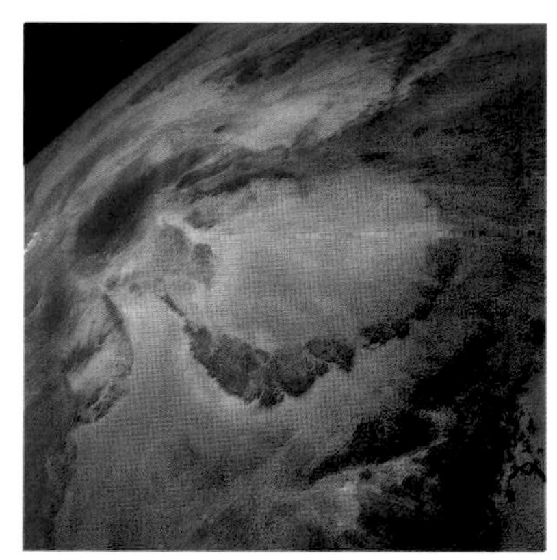

122. The Sahara Desert seen from an orbital altitude of 200 miles, showing the circular Libyan feature Idehan Marzuq. The dune formations of the irregular sandy area to its left are known as Idehan Ubari. Hasselblad 70mm transparency by Richard Gordon, Gemini 11, September 12–15, 1966.

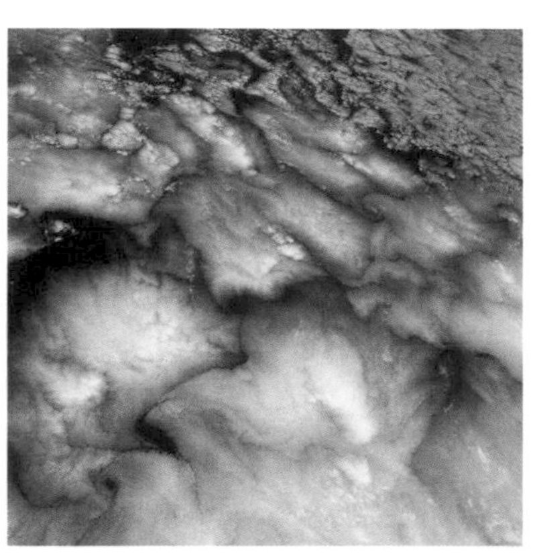

123. Stratocumulus cloud formations over the Pacific Ocean, seen from an altitude of approximately 120 miles. Hasselblad 70mm transparency probably by James McDivitt, Gemini 4, June 3–7, 1965.

124. Clouds at sunset, looking southwest over the Andes from northern Bolivia, Chile and Argentina. Hasselblad 70mm transparency by James Lovell, Gemini 7, December 4–18, 1965.

125. The three main parachutes just upon opening after reentry at an altitude of 10,000 feet, photographed looking straight upward from the command module rendezvous window. Hasselblad 70mm transparency probably by Alan Bean, Skylab 3, July 28–September 25, 1973.

126. The Pacific Ocean seen through the command module window shortly after splashdown, 230 miles southwest of San Diego, California. Hasselblad 70mm transparency probably by Alan Bean, Skylab 3, July 28–September 25, 1973.

127. An image of the full Moon seen from Earth, showing the Apollo landing sites: Apollo 11's Sea of Tranquillity, 12's Ocean of Storms, 14's Fra Mauro Highlands, 15's Hadley-Apennine region, 16's Descartes Highlands, and 17's Taurus-Littrow Valley. 36-inch telescopic photograph made at Lick Observatory, University of California, Santa Cruz, California.

128. Bombarded by meteorites for billions of years, the Moon's surface is pockmarked by craters ranging in size from microscopic pits to giant basins hundreds of miles across. Seen here is a crater measuring only a few inches in diameter. Today the Moon endures far fewer (and smaller) impacts than it did during the solar system's early history. Hasselblad 70mm negative by James Irwin, Apollo 15, July 26–August 7, 1971.

ACKNOWLEDGMENTS

This project would not have happened without the sustained and generous help of many, many people from vastly different fields and worlds. Perhaps my largest debt, however, is to the Apollo photographic archive itself, and to the astronauts who so dedicatedly recorded their missions. Right behind them are the hundreds of thousands of individuals who worked in the Apollo program and made the missions such a resounding success.

At NASA's Lyndon B. Johnson Manned Spacecraft Center in Houston, I owe a special thanks to Image Repositories Supervisor Gary Seloff and his dedicated staff of Linda Fisk and Irene Jenkins, who not only bore with the exacting demands of my image selection for months on end but made me feel warmly at home in their concrete bunker full of film. Mike Gentry and Mary-Louise Schmid did the same for me next door in the Public Affairs Library, adding their own special twist of humor. Michael McGuyer, Carlos Fontanot, and Steve Nesbitt in External Affairs, along with Rick Slater in Photography, cleared the bureaucratic path for digital scanning of the images. Chuck Welch and Sue Runco at JSC answered obscure questions about various images. Equally central thanks go to JSC's Houston neighbor The Lunar and Planetary Institute, where Head of Information and Research Services Mary Ann Hager and Photographer Debra Rueb repeatedly made their treasures and time available to me. LPI's Walter Kiefer, Jack Sevir, and Mary Noell were also helpful in lending their expertise. Tina Pauro, Librarian at the Regional Planetary Image Facility at NASA/Caltech's Jet Propulsion Laboratory in Pasadena, California, made her images and advice available in the early stages of research, as did Robert Tice, Photographic Supervisor at NASA's Goddard Spaceflight Center, Greenbelt, Maryland. Althea Washington and her staff at the NASA Broadcast and Imaging Branch in Washington, D.C., fulfilled lengthy image requests early on, and Rosemary Steinat, Librarian of the RPIF at the Smithsonian's National Air and Space Museum in Washington, D.C., provided crucial laser discs of archive imagery. Among the Apollo astronauts, the enthusiasm and support of Eugene Cernan, David Scott, Russell Schweickart, and Walter Cunningham have been invaluable. Andy Chaikin, essayist in this book and author of the definitive 1994 Apollo history, *A Man on the Moon*, has been a complete joy to work with. The generosity with which he applied his encyclopedic Apollo knowledge to the project, especially when it came to the image captions, will always be appreciated. Finally, I would like to thank Eric Jones, author of the masterfully compiled and annotated *Apollo Lunar Surface Journal* (published on the World Wide Web), whose general advice and advocacy of the project throughout the Apollo community opened many doors.

On the production side back in San Francisco, no person was more helpful in the early stages of this project than Kevin Kelly, author of *The Home Planet* (1988), a beautiful and definitive book of Earth-oriented space photography. By sharing hard-won knowlege of both the publishing and space worlds, Kevin opened the doors that made the digital aspect of this project possible, and for this I am forever grateful. Almost as invaluable was John Zax of LotusColor, who cheerfully toiled beyond any call of normal duty or reason for several months to scan about 2000 images on his drum scanner. Alexi Botkin was his able assistant, and both have become friends and trusted digital advisors. Many thanks to Jim Peck of J.P. Digital Imaging in Mountain View, California, who made his Lightjet 5000 direct-digital printer available to the project and learned just how demanding artists can be, but truly exceptional and enduring gratitude goes to Jim's master digital printer Chris Tucker, who shepherded this project as if it were his own child for more than a year. Nearest to ground zero are the talented studio assistants that have worked with me over the years: Michael Rauner, Laura Heyman, Valerie Imus, and Analucia Da Silva all tirelessly allowed me to become the proverbial eight-armed octopus. Thank you. Mark Mardon ably helped in wading through every word of the *Apollo Lunar Surface Journal*, and Jay Berry came to the rescue when I was drowning in the deep cold waters of panorama assembly. A special thanks to master digital imagist Steven Johnson, who offered his seasoned opinions early on, and to the Headlands Center for the Arts in Sausalito, California, which provided me with a room with a view for the book's final edit and sequencing. Illustrator Davi Grossman of Madcat Media in San Francisco created the book's Apollo mission diagram, and Michael Read of Elastic Studios in Emeryville orchestrated the project's presence on the World Wide Web. Various production assistances were also provided by The New Lab in San Francisco, Perrin Cutting and E-Film in Los Angeles, Chris McKinney and Digital Pond in San Francisco, Randy Greene and Muse-X Editions in Los Angeles, Steadi-Systems Video in San Francisco, Chris Crowe and Adtech Color Lab in Houston, and Evercolor Imaging in Worcester, Massachusetts.

Full Moon has been blessed with two distinguished publishing houses: Jonathan Cape in Britain and Knopf in America. I am tremendously grateful to many people for their commitment to the project. Behind them all, however, is one man based in London at Jonathan Cape, who has taken this book and made it part of his life for the last two years. Mark Holborn is that rare person who is able to generate another set of the artist's eyes and use them not only as an editor but as an actual creator. Mark has only liberated me to make this book more my own, never restricted me, and for this feat of collaboration I am eternally grateful. Simon Master at Random House UK created the transatlantic publishing structure and Sonny Mehta of Knopf has driven this project magnificently. A giant thanks to Susan Ralston, my editor at Knopf, for her sharp insight, enthusiasm, and efficiency, and to her colleagues Patricia Johnson, Paul Kozlowski, William Loverd, Kathryn Zuckerman and Nicholas Latimer for their diligent help in getting the project out into the world. Neil Bradford, production director at Jonathan Cape, has been great as has Caz Hildebrand, the art director. Applause to master printers Massimo Pizzi, Bruno Nicolis, and Paolo Scaramuzza of Amilcare Pizzi in Milan for their comprehension of *luna nero* and all that it has entailed.

Sandra Phillips, Curator of Photography at the San Francisco Museum of Modern Art, has long supported my work, and for her early and extensive commitment to an exhibition of *Full Moon* at SFMOMA in Autumn 1999 I thank her profusely. Associate Curator Douglas Nickel has offered valuable insight, as has Graphic Study Manager Tom Sempre. Able assistants Ariadne Rosas and Marceline Trujillo have kept it all together, and thanks to Exhibitions Director Barbara Levine for her help. The enthusiasm of Dennis Davidson, Astronomical Artist at the American Museum of Natural History in New York, hooked me up with Neil Tyson, Director of its Hayden Planetarium. With the help of Traveling Exhibits Coordinator Rose Wadsworth, I thank them for considering an inaugural *Full Moon* exhibition at the Museum's spectacular new Rose Center in early 2000. In London at the Hayward Gallery, I am especially grateful to Director Susan Ferleger Brades and Head of Exhibitions Martin Caiger-Smith for their enthusiastic commitment to a show of *Full Moon* in the summer of 1999. Many curators have taken supporting and advisorial roles in the project, among them Merry Foresta at the National Museum of American Art in Washington, D.C., Jock Reynolds at the Addison Gallery of American Art in Andover, Massachusetts, Deborah Klotchko at the Friends of Photography in San Francisco, and May Castleberry at the Whitney Museum in New York. Special thanks to Andy Grundberg for advising me early on, and Christopher Phillips at *Art In America* for his keen lunar enthusiasm. Craig Krull of Craig Krull Gallery in Los Angeles has been wonderful. Catherine Edelman of Catherine Edelman Gallery in Chicago has been a real support, as has Thom Healy of Thomas Healy Gallery in New York.

Perhaps most impossible to adequately thank are those friends and family who have repeatedly given so much to me and the project over the last four years. To name only a few, Christopher Sharpe, Paul Catasus, David and Lisa Rattray, Deborah Light, The Burning Man Clan, Michael Rauner, Pauline Shaver, Jack Woody, Thomas Long, Pete Joel, Laura Heyman, Larry Sultan, Jim Goldberg, Sven Wiederholt, Dara Freed, Niels Thorsen, James Crump and Arnold Kemp have all been wonderful, in countless different ways. A special thanks here to Jerome Oremland. Finally, I owe a great creative debt to Al Reinert, whose 1989 documentary Apollo film, *For All Mankind,* served as an early inspiration for *Full Moon*. Al was also selflessly generous in helping me obtain, from the master positive stored away in his hall closet, the digital stills that comprise the fiery beginning of this book.

ALL PHOTOGRAPHS COURTESY NATIONAL AERONAUTICS AND SPACE ADMINISTRATION

DIGITAL SCANNING AND IMAGE PROCESSING BY MICHAEL LIGHT STUDIO

EDITED WITH THE ASSISTANCE OF MARK HOLBORN

PRODUCTION AND DIRECTION BY MARK HOLBORN AND NEIL BRADFORD

DESIGN BY MICHAEL LIGHT

Published by Jonathan Cape 1999

1 3 5 7 9 10 8 6 4 2

Telescopic image of the Moon reproduced with permission of UCO/Lick Observatory
Orbital diagram by Davi Grossman

The research cited and observations expressed in *Full Moon* are those of the authors,
not those of the National Aeronautics and Space Administration

First published in Great Britain in 1999 by Jonathan Cape
Random House, 20 Vauxhall Bridge Road, London SW1V 2SA

Random House UK Limited Reg. No. 954009

A CIP catalogue record for this book is available from the British Library

ISBN 0-224-05128-8

Printed in Italy by Amilcare Pizzi S.p.A, Milan

+ + +